STOREY'S GUIDE TO

RAISING LLAMAS

STOREY'S GUIDE TO
RAISING
LLAMAS

Gale Birutta

Storey Publishing

Grateful acknowledgment is made to Lynn Lenker and Tim Barrus, who generously made their farm, Lithia Llamas, available for our photographer.

The mission of Storey Publishing is to serve our customers by publishing practical information that encourages personal independence in harmony with the environment.

Edited by Alice Vigliani and Suzanne Donahue
Cover design by Susan Bernier
Cover photograph by Susan Ley
Back cover photograph by PhotoDisc
Text design by Mark Tomasi
Production by Mark Tomasi, Allyson L. Hayes and Erin Lincourt
Line drawings by Becky Turner, except for pp. 107 and 112 by Elayne Sears
Indexed by Susan Olason, Indexes & Knowledge Maps

The information in this book is true and complete to the best of our knowledge. All recommendations are made without guarantee on the part of the author or Storey Publishing. The author and publisher disclaim any liability in connection with the use of this information. For additional information please contact Storey Publishing, 210 MASS MoCA Way, North Adams, MA 01247.

Storey books are available for special premium and promotional uses and for customized editions. For further information, please call 1-800-793-9396.

Printed in the United States by Versa Press
10 9 8 7

Library of Congress Cataloging-in-Publication Data

Birutta, Gale, 1951–
    [Guide to raising llamas]
    Storey's guide to raising llamas / Gale Birutta
         p. cm.
    Originally published: Guide to raising llamas. Pownal, Vt. : Storey Communications, c1997.
    Includes index.
    ISBN 1-58017-328-4 (alk. paper)
    1. Llamas. I. Title.

SF401.L6 B57 2000
636.2'966—dc21
                                        00-030803

CONTENTS

FOREWORD

The llama industry has enjoyed considerable growth in the past two decades. With this influx of new llama owners comes the need for a practical guide. Most books on the subject of llamas assume that you already own one; they don't consider the many people who want to know how to get started with these delightful animals — and what to do with them once they've brought them home. That's where this book comes in.

This guide will take you from your first steps in acquiring a llama through all aspects of its care. You'll discover numerous business opportunities and learn how to manage, market, and expand your enterprise. If you want to breed llamas, you'll learn how to recognize a top sire and where to advertise his services.

I'd like to extend a welcome to the wonderful world of llamas. These playful animals will enchant you with their antics. Their docile, tranquil nature will ease away your tensions. Llamas provide a host of business opportunities whether your interest lies in breeding, fiber production, pack expeditions, or even marketing fertilizer. Llamas are excellent guard animals that stand sentinel over your other livestock. They make terrific 4-H projects. As companions, their therapeutic benefits are well known. And if you keep llamas as pets, well, your children will pass countless hours enjoying them (and so will you).

It's easy to fall in love with these remarkable animals: Who could resist those enormous, trusting eyes? Once you fall under their spell, you'll be entranced for life, just as I am.

PART ONE

GETTING STARTED

ONE

INTRODUCTION TO LLAMAS

There are over 95,000 llamas in the United States today. This dignified and appealing animal is valued for its agreeable disposition, high-quality fleece, and capacity to be trained as a work animal. Llamas are wonderful companions as well as productive livestock. They can reach the age of 25. But what exactly are llamas, and where did they come from?

Background and Heritage

Llama ancestors existed on the plains of North America approximately 40 million years ago. One theory holds that during the Ice Age, these llama ancestors were forced to move into South America. Some of them migrated north across the Bering Sea land bridge to Asia, evolving into the camel of today. Those that migrated south evolved into two distinct species, the wild *vicuna* and the *guanaco*.

A popular misconception is that llamas are native to South America, and some people think their origins are in Asia. In fact, llama ancestors could be found in *North America* some 40 million years ago!

The Vicuna

The vicuna thrives easily above 17,000 feet in the Alti Plano of Peru. The vicuna is valued for its outstanding fleece, which is lustrous, warm, and easy to weave into beautiful coats, blankets, and hats.

Vicuna mothers usually give birth during the morning hours, so that the newborn will be able to dry off in the warmest part of the day. Babies born later do not have time to dry off and consequently may freeze to death. Vicuna babies must be able to run with the herd within 15 minutes of birth. Unfortunately, 30 to 40 percent of vicuna babies die.

Vicuna

In South America, the four camelid cousins (vicuna, guanaco, llama, and alpaca) are termed "lamas" (pronounced *yamas*).

The Incas protected the vicuna and its valuable fleece. Vicuna fleece was so highly prized that only royalty could wear it; any others caught wearing vicuna were put to death. Yielding one of the finest fibers in the world, the vicuna was indiscriminately slaughtered for its hide and fleece by the Spanish in the 1500s.

Nearing extinction (fewer than 2,000 animals remained), vicunas were declared an endangered species in 1969. Through the efforts of Dr. William Franklin of Iowa State University, the Pampa Galeras Vicuna Reserve was established on the western edge of the Andes Mountains in Peru. By 1990, only 20 years after the establishment of the reserve, the number of vicunas increased to 30,000. Today, approximately 200,000 vicunas make their home in Pampa Galeras.

The South American government has banned trade of products made from vicuna fiber, but that has not stopped demand. The reserve constantly battles poachers, who slaughter hundreds of vicunas each year. Today, the South American government allows Indians to harvest fiber from any vicuna that crosses their property. Vicuna pelts bring as much as $1,000 on the black market. Any vicuna fiber found in the United States has almost certainly been purchased on the black market. Some products labeled vicuna may be an imitation, such as baby camel.

Classification

Class	Mammalia
Order	Artiodactyla
Sub-order	Tylopoda
Family	Camelidae
Genus	Lama
Species	Glama

The Guanaco

Guanacos are the wild cousins of the domestic llama and alpaca, as well as the vicuna.

The guanaco is somewhat leaner in build than the llama and has a unique reddish-brown coloring with a white underbelly and white fleece on the inside of the legs. The guanaco's range lies from northern Peru to the southernmost tip of Chile and from sea level to 12,000 feet. The largest numbers roam the Torres del Paine National Park in Chile.

Under protection of the state, guanacos now number over 1 million. Their only natural predator is the puma, which kills up to 10 percent of all guanaco babies. Some guanacos have been imported to the United States with llamas over the past several years (there are approximately 150 guanacos in the

United States as of December 1996), and guanaco fiber is just now finding its niche with fiber producers.

Guanaco

The Domestic Llama and the Alpaca

The domestic llama and the alpaca — the only two species of camelids to be domesticated — were developed through thousands of years of controlled breeding by the Incas. The Incas used llamas as beasts of burden as well as for meat, hide, and sinew; the woolly alpaca was bred for its fine fiber.

Alpacas may be the oldest domesticated animals in the world. Most of them today make the Alti Plano in Peru their home, living at altitudes of 14,000 to 16,000 feet. The fine fleeces from these animals provided warmth to the Andean Indians and protection against the harsh Andean environment.

The first alpacas were imported into North America in 1984 from Bolivia and Chile. Peru considers the alpaca its national treasure, only allowing export of these animals into the United States since 1993. Currently, there are 10,000 alpacas in the United States.

There are two types of alpacas. Wool of the Huacaya alpaca possesses crimp, making it easy to spin. The Suri alpaca is rare, making up only 10 percent of the world's alpaca population. Its fleece is extremely fine with very little crimp. The fiber hangs from the animal in locks.

Kevin Kennefick

Domestic llama

Alpacas seem to have few birthing problems. It is rare that an alpaca mother rejects her cria (offspring). An alpaca baby usually stays close to his mother's side. Alpaca mothers and crias show a strong bonding instinct. Llama crias, generally more independent, often stray from their mothers. Both alpaca and llama mothers have been observed nursing another female's cria.

North American Llamas

Llamas were imported to this country in the late 1800s to various zoos nationwide. William Randolf Hearst of California and Roland Lindemann of the Catskill Game Farm in New York simultaneously developed the first commercial llama herds in 1928. In 1930 the U.S. Department of Agriculture imposed an import ban on camelids from South America, for fear of foot-and-mouth disease. Today, camelids from Peru and Bolivia are considered at high risk for foot-and-mouth disease. They must spend 60 days in quarantine before they leave South America, and then they spend an additional 90 days at the Harry S. Truman Quarantine Center in the Florida Keys. Chile has been deemed free of foot-and-mouth disease, and camelids entering from that country are required to spend just a 30-day quarantine in New York.

Alpacas fare well in the harsh Alti Plano in Peru.

The Catskill Legacy

After the death of William Randolf Hearst in 1956, Roland Lindemann purchased the Hearst llamas and shipped them to his Catskill Game Farm in New York. The Catskill Game Farm soon became the premier importer and breeder of llamas in North America, a position it still holds today. Hearst and Lindemann were the only known breeders to import llamas. These men were the leaders in the industry's early stages.

Earliest records show a first North American llama birth in the 1870s, with approximately 30 zoos and 18 private individuals

Nola Graham

The Suri alpaca boasts an extremely fine fleece with little crimp.

owning or breeding llamas prior to the 1930 ban. Between 1878 and 1928, one eastern zoo reported over 40 births.

The Catskill Game Farm selectively bred the first large herd and developed most of today's top foundation domestic bloodlines. "Foundation" bloodlines are those from which most of the modern-day generations of llamas are derived. Compared with their South American cousins, Catskill llamas are typically smaller, stockier, and more compact; they also have shorter necks, distinctive "banana ears," and beautifully proportioned heads with sharp, chiseled features. Descendants of the Catskill foundation llamas have their own distinct "look" and can be quickly identified in any herd.

In the past few decades, llamas have been imported from South America through the quarantine stations. They have also been imported from New Zealand and England. As the original gene pool in the United States dwindles, it becomes necessary to import new blood. Proper diet and good veterinary care have contributed to making today's North American llamas healthier and larger than their South American counterparts.

A Short-Lived Fad

In the 1970s llamas were purchased for fiber and as pack animals. Many new owners had no previous livestock experience; nonetheless, they believed they would recoup their investment with breeding stock. It became trendy to own llamas. With so many eager new buyers, the demand quickly outstripped the supply. All of a sudden there were long waiting lists for animals and prices rose to extremes.

Prices for female llamas started at $10,000, with no consideration given to conformation (proper proportionate bone structure), fiber quality, or reproductive history. Only a very small percentage of breeders actually bred for specific characteristics; most of these individuals bred strictly to produce female offspring.

The skyrocketing prices and quick profits attracted wealthy investors. Auctions drove prices higher, with no apparent ceiling. It was not unusual for a highly promoted male llama to sell for more than $150,000. But then, the abrupt dive in the economy in the mid- to late 1980s had its effect on the llama industry.

In March 1989, a group of get-rich-quick investors attempted to import llamas and alpacas from South America. Trying to keep their profit margins high, these American investors left Chile with 278 llamas and alpacas, although the ban on importation was still in effect. To get around

Montpelier

Montpelier — Monte — was the second llama we purchased, back in 1988. We visited the farm where Yamaha, as he was called then, was guarding the dairy goats. (These folks were selling their llamas, as they owned a profitable dairy goat product enterprise and wanted to devote all their time to it.) When we arrived, the owner informed us that the llama was "Catskill," originally purchased from the game farm in 1979 as part of a breeding pair. "Catskill!" We could hardly contain ourselves. At this time Catskill stock was selling in the five-digit range. We quickly got out our checkbook.

We were unable to get near him. If ever there was a wild llama, he was it; he simply was not catchable. The owner agreed to deliver him the following weekend. Yamaha, later renamed Monte, arrived full of spit and holler.

Montpelier, the author's senior herd sire, was originally a member of the 1979 Catskill Game Farm herd.

Gale Birutta

Barely off the trailer, he looked me straight in the eye and spit squarely between my eyes. (Llamas rarely spit at people.) Oh well, I thought, he's upset, he'll calm down. Monte did not calm down; he screamed violently and spit at the slightest touch.

Monte continued to be difficult to handle, enough so that we thought we had made a mistake not only in buying this male, but also in getting involved with llamas in general.

Then along came the female, Jamaika, a year later. Monte had lost his mate many years earlier on the goat farm and was lonely for a lady. Well, he became a new man. Monte is now an angel to work with and produces some of the finest offspring around.

government red tape and to avoid the Harry S. Truman Quarantine Center, their plan was to quarantine the animals in Antigua. The plan went haywire. A feud arose among one of the American investors, the local people, and the Antiguan government. The animals were not allowed to enter the country and were banished to a scorching, barren desert island off Antigua's coast.

These investors were not camelid breeders; one was a Russian wolfhound breeder from Texas. This man contracted for 160 animals and reportedly already had 49 of them sold for $500,000. One female llama had a buyer in the United States who would pay $80,000.

Most of the camelids died from the heat, more than a third of them less than a month after their arrival on the island. After several years, only a handful of these llamas and alpacas remained, now moved to and cared for on the main island of Antigua.

Gale Birutta

Llamas are intelligent and alert companions.

Stabilization and a Growing Market

With the investors gone, new breeders began popping up. Many had very little experience in herd management, feeding and nutrition, veterinary needs, or llama reproduction. Information was lacking, so these new breeders sought the advice of those from whom they purchased their llamas. Networks grew, associations formed, and education in the llama industry was off to a running start. The International Llama Association (ILA) was established in January 1982 to promote the sharing of information and to help educate new and potential owners. As the ILA grew, regional chapters developed. Today, the ILA holds an annual conference, provides educational workshops, and prints educational brochures, a membership directory, and a newsletter. It also maintains a hot line to access veterinarians. Research and education — the main objectives of the ILA — filter down through the regional chapters.

Serious breeders began defining and setting goals in their programs. They paid close attention to breeding for fleece quality, length, and density, and for heavy bones and frames to carry the heavier fleeces. Breeders of packers sought to breed taller, less bulky llamas for moving out quickly on the trail. Some bred smaller llamas to appeal to people intimidated by larger livestock. Everyone found his niche, and today there are llamas to meet everyone's desires and needs. Prices are stable and the llama market is larger and stronger than ever before. With the Eastern Seaboard still commanding the highest prices, the average price range today is between $500 – $2500 for males and $1500 – $4500 for females. Prices in the Midwest and on the West Coast can be as much as one-third to one-half less. Although breeders and buyers still expect a profit on their investment, they now realize that good management is an integral part of producing top-quality animals. People are purchasing llamas for the genuine joy of raising them, not simply with profits in mind. Investment returns are inevitable, but appreciation for these animals continues to grow.

Llama Uses Today

Llamas, originally bred as pack animals, carried burdens for the Indians of South America for centuries. Even today, poor Andean people still rely on llamas to carry to the marketplace fruits, vegetables, and other items for trade. Andean sheep breeders rely on their llamas to carry supplies while they move sheep herds to new browsing grounds. In the United States, llamas have adapted well to their primary role as a recreational animal, but owners and breeders are finding other reasons to keep llamas.

Packing

The number of commercial packers and recreational hikers who use llamas is on the rise, and llamas are finding their way into backyards around the country for use as "personal packers."

After 5,000 years as a packing companion in the Andes, the llama is the ideal candidate for commercial outfitters, campers, and recreational hikers and walkers. In addition to hauling extra gear, they provide quiet, stable, environmentally friendly companionship on the trail. High temperatures can limit the heavy packing of llamas, especially in the Southeast and Southwest. Only in the higher elevations of the Southwest are llamas comfortable with the heat. Terrain poses no physical restrictions to llamas, but some state and federal regulations prohibit pack animals in certain areas. However, on some federal lands, horses are banned but llamas are permitted.

Recreational hiking with llamas is becoming increasingly popular.

Llamas used as packers offer many benefits. Active retirees are among those who now own personal packers. For energetic hikers who for any number of reasons are now unable to carry gear, personal pack llamas provide the opportunity to continue to enjoy outdoor life and the environment.

Backcountry fishermen can easily load their equipment onto the backs of llamas and reach fishing areas that would be otherwise inaccessible. Llamas can carry inflatable rubber boats that anglers can use for fishing on the water.

Surveyors find llamas useful on the trail, again in areas that are inaccessible by conventional means. Carrying surveying equipment, tools for trail clearing, and meals, llamas are an alternative to horses and four-wheel-drive vehicles — and will not harm delicate ground.

Llamas adapt easily to any outdoor activity that involves hiking or walking with the need to carry a heavy pack and equipment.

Dennis Jensen

Kevin Kennefick

By shearing your llama once per year, you can recover the cost of its upkeep by selling the fleece.

Fiber

Although not originally bred as fleece producers, llamas do provide their owners with the excellent by-product of high-quality fleece. In fact, interest in llama fiber has swelled in the 1990s. Now llamas are also bred for high-quality fleece for hand spinning, felting, and fiber used in crafts. With an established market, their fiber is increasingly popular with sheep producers. Mixed with wool from sheep, llama wool adds luster and sheen. Llama fiber does not shrink, and is lightweight and warm.

Llama fiber is naturally hypoallergenic (it has no lanolin) and can be used to stuff pillows. It's a blessing for people who are allergic to other natural fibers. During the Christmas season, crafters use white llama fiber for Santa's beards. It is not necessary to shear all llamas, but warmer climates tend to dictate shearing of llamas at least once per year.

Fertilizer

Llamas are amazingly clean. Their communal dung pile makes clean-up easier than it is with other livestock. The manure is in the form of pellets — odorless — similar to that of deer or goats. High in nitrogen, this manure

Guardian Llama Placement Program

Llamas make effective livestock guardians when properly screened.

can be used on gardens without the danger of burning plants. Composting is not necessary, but will enhance the naturally high nitrogen content of the manure.

Harvesting manure is a commercial fertilizer venture for many llama owners, and it can be highly profitable. With the emphasis on the environment, natural fertilizers are much in demand. Developing a specialty fertilizer from llama manure can yield as much as $10 for a 2-pound bag, similar to the cost of bat guano. The quality of llama manure–based fertilizer can also be compared to that of bat guano, which has an average rating of 4-3-2 (ratio of nitrogen-phosphorus-potassium).

Livestock Guardians

Llamas are highly territorial. If they are permitted to grow up and socialize with their own species, at 18 to 24 months of age you can remove them from the llama herd and put them to pasture with other livestock. Llamas used as livestock guardians have saved ranchers and farmers thousands of dollars in losses to predators. Not all llamas will become good guard animals; however, research continues and specialty guardian llamas are now appearing all over the United States. Llamas now guard sheep, goats, ostriches, dairy cows, fallow deer (domestic Oriental deer raised for meat), and even poultry.

Cart Driving

The natural sensitivity of the llama's head makes the llama an excellent candidate for cart driving because it will respond readily to halter and reins. A specially designed halter with reins attached guides the llama. A single llama, pairs, or a larger team will work together to give a fun ride.

Llamas hitched to pony carts provide Sunday-afternoon rides for families. In Aspen, Colorado, they pull sleds for skiers and tourists and are used Nordic-style (with the reins held directly behind animal) for modified cross-country skiing. Llamas do not shy or spook the way horses will, and this makes them safe bets for road driving.

Upkeep

The expense for maintaining a llama is less than that of the family dog. Each llama consumes approximately a bale of hay per week and forages during the grazing months. Because llamas are highly disease-free and disease-resistant, their veterinary costs are far below that of other livestock. We estimate that you can keep six llamas for the cost of keeping one horse.

As many as five llamas may graze and forage comfortably on one or two acres of land (in contrast, one horse requires two acres). One or two llamas may be kept on a backyard lot. Llamas have soft, padded feet that are easy on pastures. They don't dig up grazing areas. A wide variety of forage appeals to them. Llamas are not normally fussy eaters; for example, they enjoy grass and berry bushes. Their grain consumption is minimal, depending on the age of the animal and whether the llama is used for breeding. (See more regarding nutrition and feeding in Chapter 3.)

Manageability

Llamas adjust quickly to new situations. A conscientious breeder will help a customer select the llama best suited for him and will help that new owner for as long as necessary. Highly intelligent, llamas tend to take advantage of inexperienced owners (much as horses will). Llamas will play "catch me if you can" if you let them, but they will quickly learn who is boss. Handle a new llama with slow, easy, and deliberate movements.

Although llamas are not riding animals, small children are light enough for most any llama to carry.

Other livestock may react in fear when they first meet llamas. Horses are wary and will stomp and snort at their new neighbors; the llama will stand quietly and aloof as if to ask, "What's your problem?" Generally, though, other livestock will accept llamas within a couple of weeks. The adaptable llama settles in quickly, usually within several hours. I have seen llamas that have traveled several thousand miles, endured handling by veterinarians and many others, been shipped off to two or three different places for housing before reaching their final destination, that were calm and settled in within an hour or two.

Introducing llamas to other llamas, however, requires some precautions, particularly if the new llama is an intact (nongelded) male, an open (nonpregnant) female, or a weanling. Generally, an intact male — new to a herd or not — should not be housed with another intact male, especially if there are females around. (See Chapter 4 for more on housing.) Geldings can usually be introduced to other llamas with no problems, and are the easiest to handle. This is one of the reasons why geldings are usually recommended for the new owner.

Because of its basically docile nature, a llama is generally safe for children to handle. Llamas are the only livestock species for which juveniles are allowed to exhibit an intact male in the show ring. Llamas generally do not kick, bite, or step on people. They are naturally aloof and would rather keep their distance, which gives them a dignified presence.

Llamas are easy to train and usually learn in just two or three repetitions. You can transport them in a station wagon or the family minivan, train them to jump in the back of a pickup, and even travel with them in air boats, light planes, canoes, and gondolas!

Llamas are respectful of fencing; you may confine them in most types. If you use electric fencing, there is no set number of strands. Many llamas will stay put with as little as one strand of wire. When housing breeding animals, though, use a more substantial electric fence, one with three to five strands. Roll-up rotational sheep fencing may be adequate, particularly when animals are moved frequently. Housing needs are minimal; llamas require only a three-sided shelter from winter wind and hot summer sun.

Because they are inexpensive to keep, are manageable, and require minimal housing, llamas are finding their way into dairy operations, horse farms, and sheep and goat farms. The llama's natural calming effect and guarding instinct make it a welcome addition to a herd or flock. Llamas as guardians may warn off coyotes while cows are giving birth, deter neighborhood dogs from bothering show horses, and save thousands of lambs and kids from falling prey to various canine species.

Two

Buying Your First Llama

Once you've been bitten by the llama bug, it is difficult to restrain yourself from rushing out to buy that first llama. But be wise: First join a llama organization in your region and educate yourself. Management and care will depend on the type of llama you want and what you will use it for. With so many options, you must be educated enough to make an intelligent choice.

Look at color, quality of fleece, reproductive capability, and structure if you seek llamas for breeding stock. Llamas chosen as pack animals should be tall and leggy. Pets must have a good disposition above all else. Visit as many llama farms as possible; you'll get an excellent indication of what is available. Ask each breeder to refer you to other farms in the area that may have what you are looking for.

Starting Small

Start small; consider beginning with just one llama. If you're not sure, one inexpensive gelding (castrated male) will give you good hands-on experience. If you're committed to owning llamas, try to purchase a pair. Llamas are highly social animals and need to interact with other livestock. When kept with other llamas or other livestock, llamas are amusing, nurturing, and friendly. When kept alone, llamas may become frustrated and aggressive toward humans.

A single new llama, whether it is a gelding, intact (nongelded) male, or female, can generally be housed with most other livestock. An intact male should not be housed with female sheep or goats. The new llama will adapt well, usually within a few hours. In the case of a pair of llamas, geldings can be kept with young llamas, females, or — depending on the aggressiveness of the intact male — an intact male. You can often place an intact male with bred females, but you'll need to remove him at the time of delivery.

A starter llama will allow you to get a feel for the animal, examine the market, define your needs, and set some goals. Don't be in too much of a hurry to purchase a lot of animals; take your time and look for the best animal for the money.

In almost all cases, you'll keep the starter llama while your herd grows. This llama will become a favorite. If chosen correctly, he'll be the one who acts as a companion for all new llamas and will share quarters with them. A starter llama will also help you sharpen your skills in handling and training.

Starting small but with quality animals will reap rewards in the end. I've been a llama owner for 10 years, and my herd hovers around 15. Five are packers, the rest are breeding stock. When I entered the market in 1987 with one llama, other breeders were purchasing 5 to 10 females each to get a quick start (prices at that time were $8,000 to $10,000 apiece for females). Today, some of these same breeders are producing inferior animals because they purchased hastily without defining their needs and setting breeding goals.

How We Got Started

Once you own and are around llamas on a daily basis, it is hard to imagine life without them. When John and I learned that llamas were being raised as domestic livestock at Cathedral Pines Farm in Connecticut, I immediately contacted these folks and arranged a visit. We spent a glorious afternoon with the wonder-

Gale Birutta

ful Calhoun family and these mystical creatures. That was it. It was decided just like that. We made plans: We would get married, move to Vermont, and raise llamas. We decided to play on the "Made in Vermont" slogan and named our breeding business "Made in Vermont Llamas." All llamas used for breeding or born on our farm would be named after Vermont towns. So far we've raised Bristol, Cabot, Sheffield, and many others.

Although Bennington was our first llama, his love for horses eventually sent him to a new owner as a horse guard.

Defining Your Needs

Carefully think about what your intentions are. Do you want to become a serious breeder? Will your llama be a weekend hiking or camping companion? Will you use him as a 4-H project for the kids? Do you need a livestock guardian? Would you like him as a pet? Figure out what you want and learn what type of llama will fit that bill. Sit down and establish written goals covering at least the next two years. You'll be able to refer to them often. Adjust your sights periodically.

Goals

	First Year	Second Year
Pet	Learn basic care Learn basic handling	Extend handling and training
4-H Project	Introduce llamas to kids Teach basic care Teach handling	Teach showing Teach shearing
Hiking/Camping Companion	Begin daily routine of conditioning yourself and llama Begin training to load Purchase initial equipment for packing/hiking	Further training to load in different types of vehicles Experiment with different types of loads Extend length of trips
Livestock Guardian	Study personalities and attitudes of your potential guard llamas Plan to promote and market your guardians through media/public speaking	Fine tune your screening process Expand regions where you are placing your guardians
Serious Breeder	Purchase foundation stock, preferably bred female Decide what type of stock you would like to breed: classic, show, etc. Attend shows and take a show clinic	Scrutinize offspring Cull stock if necessary Research and locate outside stud for rebreeding, or purchase stud Begin showing

Farm and Ranch Visits

Almost every state or province in North America has a llama association, and there are many regional and national associations as well. These associations can supply the names of breeders in your area. (A list of these associations can be found in Appendix A.) By visiting various farms, you will be able to see llamas that are more typical of the norm; at shows and auctions you'll see only the breeders' very best. Farm visits also give you the opportunity to see parents, siblings, and offspring. Purchase close to home if you can — it's more likely you'll be able to get continuing support from a local seller.

After reading all the educational material you can get your hands on, ask a breeder to assist you in making your selection. An honest and helpful breeder will be happy to help you buy from another farm, should she not have what you are looking for.

Deal only with established, reputable farms. Do not buy on impulse at the first farm you visit; it is unlikely that you will find the animal you are looking for right away. Plan on visiting at least 10 farms to get a feel for the llamas available within your budget. Look around the breeder's farm for general cleanliness and proper herd management. This is an excellent indication of how the breeder cares for her animals and runs her operation. Finally, ask the farm for references from people who have recently purchased llamas there. This will give you feedback regarding the breeder's reputation and credibility.

At each of the farms, the breeder should offer you a halter and allow you to catch, lead, and even groom your potential llama. Any reluctance on the part of the breeder to allow you to catch, halter, and handle the llama should send up a red flag.

All llamas, no matter what their age or what they will be used for, should be halter broken and trained to lead. Don't even think about a llama that is not halter broken. If you are considering a weanling llama that needs to be halter broken, tell the breeder that you require the youngster to be halter broken before you take ownership.

A reliable breeder may ask you as many questions as you ask her. Five of the most common are:

1. Have you had any previous livestock experience?
2. What type of facilities do you have or will you have for this llama?
3. What do you plan to use this llama for?
4. Do you have any other livestock, pets, or children?
5. What is your price range?

These questions will show that you are in fact dealing with a compassionate and caring breeder who takes pride in where her llamas go and how they will be used and treated. Also, be sure to ask if the breeder provides a purchase contract. If not, go elsewhere. A purchase contract is a written agreement between the buyer and seller, explaining the terms of the agreement such as payment and guarantees, if any.

A reputable breeder will offer a health guarantee in writing without any questions. As the purchaser, you can supply the seller with a contract outlining your requirements. When you feel you have found the ideal llama, then you and the seller should agree on a price, delivery date and method, and other terms, and then sign the contract. (See sample purchase and sales agreement on page 31.)

Selecting the Best Animals

As you visit farms and ranches, keep in mind your written goals and ask yourself: "Can this llama meet my needs?" A llama must meet *your* definition of the ideal llama. Measure any new animal against other animals you have seen.

Compare all llamas as you would other livestock for conformation (proper proportionate bone structure), gait, and way of moving. (For more on conformation, see Chapter 5.) Disposition and personality are important if you are selecting a pet or 4-H project for the kids. An unruly llama should be avoided. For breeding, the overall balance of the animal is important. Llamas that are selected solely on aesthetics and that lack sound structure are quite likely to produce conformationally weak offspring. Don't be mesmerized by flashy animals. You can always breed beautiful fleece. It is often best to opt for the taller, heavier-boned animal with excellent conformation.

Because the llama industry is still in its infancy, type distinctions in breeds of llamas have not been established. It is best to consider *types* of llamas: *standard* (having light to medium wool) and *heavy*

Gale Birutta

A heavy-wool llama with "pantaloons."

wool (with abundant fleece). A pack llama usually falls into the classification of standard, but a heavy-wool llama can be a packer if it is shorn. (See Chapter 13 for more on fleece.)

There is less risk involved in meeting your expectations when you choose a proven animal. A "proven" llama is a llama that has proved itself either as a breeder through having produced offspring, a packer that has experience, or perhaps as a reputable, experienced guard animal. Proven simply means experienced. The unproven llama is a llama that has not reproduced as yet, has not previously packed, or has never guarded livestock. The unproven llama will be a greater risk but will offer you a monetary savings. The decision is yours.

Criteria for Selecting Your First Llama

Intended Use	Options	Considerations
Fiber	Light to medium wool	Staple length Color Density Coverage Crimp/curl Fineness
	Heavy wool	Color Density Coverage Crimp/curl Fineness Is llama accustomed to being groomed or shorn?
Packing	Commercial use	Should be pack trained Tall, correct, and moves out readily Must get along with other llamas in string Weight limits Miles logged Type of terrain covered Leader or follower Water trained
	Recreational use	Tall, correct, and moves out readily Pleasant personality Willingness to get along with other llamas Easy to work around Leader or follower Water trained

Intended Use	Options	Considerations
Breeding	Stud purchase	Proven breeder Easy to manage and handle Registered and blood-typed Reproductive history Veterinary history Winnings, if any
	Female purchase	Matron female Registered, preferably blood-typed Reproductive history Veterinary history
	Outside stud service	Same as for stud purchase View offspring Show winnings, if any
Pets, **Companions,** **and 4-H**	Weanling llama	Halter trained Lead trained Good disposition Must be accustomed to handling and 　easy to work around
	Aged llama	Gelding Halter/lead trained Good disposition Must be accustomed to handling and 　easy to work around
Guarding	Single guardian	Free-range Must be completely bonded to specific 　livestock; cannot stray Gelding 18 months and older Previously housed with type of livestock 　it is assigned to protect Respectful of fencing
	Coguardian	Must be used to and will accept dogs 　as coguardians Gelding 18 months and older Previously housed with type of livestock 　it is assigned to protect Respectful of fencing

Breeding Stock

If you're shopping for breeding stock, ask:

- ◆ Is this llama registered and blood-typed?
- ◆ Has this llama sired offspring or given birth previously?
- ◆ How many progeny does this llama have?
- ◆ Were the deliveries normal and the crias healthy?
- ◆ Where may one see offspring or photos of them?

Breeding stock will be your largest financial investment and will require the most research on your part. Being well informed and clear will help you to choose the background, bloodline, or traits you intend to produce. If you are a member of the major llama associations and subscribe to some publications, you will eventually receive catalogs for llama sales and auctions. Prospective buyers would be wise first to join their regional llama associations (regional associations are listed in Appendix A). Regional associations publish newsletters with information regarding breeders in your area, as well as shows and fairs to attend where you can personally speak with these breeders. Several publications are listed in Appendix C; these are all excellent resource materials and contain many up-to-date research articles on breeding and management. They are also excellent reference materials for the study of bloodlines and for comparing photos of types of llamas, and are a general help in introducing you to "types" or "styles" of breeding llamas.

While auction sale catalogs provide lineage information on studs and females, be aware that some breeders do not provide untouched photos of their animals. It is common knowledge that many photos are airbrushed to enhance the llama's looks. In fact, even a videotape can be altered to improve the animal's looks. When considering a particular llama, make sure you inspect the llama personally.

When considering a young animal of weaning age (5–6 months) who has several faults (conformation, temperament, etc.) understand that while he may outgrow them, it is also possible that he won't outgrow them and that he will pass these faults on to his offspring. The most common fault in young llamas is crooked front legs. Ears may also be straight and short without the more desired "banana" shape. Upon maturity, less desirable ears may grow into longer, more appropriately curved ears. This is where your research comes into play. If it is apparent that these faults do correct themselves with maturity, then this unproven male or maiden female may be a good buy. (A full sibling with no visible faults will command a much higher price.)

When evaluating a young animal for purchase, inspect the animals' parents — this will give you a clearer picture of what the youngster will look like as an adult. Ask to see mature siblings and half-siblings. Check them for straightness of back, tailset, and correctness of legs and ears.

Registry

If you are interested in breeding, it is crucial that the llama you buy is registered with the International Lama Registry (ILR). Ask to examine the ILR registration papers. These carry a three-generation ancestral history; a five-generation ancestry is available upon request. The registration papers reveal any line breeding. (When a male and female llama are mated together so their offspring remains closely related to one desired ancestor, this is called *linebreeding*. See Chapter 9.)

The registry is now closed; therefore, if the parents of the offspring are not registered or are not registerable, the offspring cannot be. This unregisterable animal is virtually useless in a breeding program because your buyers are not likely to purchase an unregistered animal. Ask the breeder to show you the registry certificate for each animal you are interested in. Even if the registration is pending, there's no guarantee that it will pass muster. Make the final sale contingent on a bona fide registration certificate. (See Chapter 6 for a sample registration certificate.)

As of October 1996, there were 94,320 llamas registered in the United States, 996 listed (unregisterable), 175 screened imports, and 186 at some point in the interim screening process. It's impossible to estimate how many llamas there are that are *not* registered or listed with the ILR.

A registered llama is not mandatory as a companion, for fleece production, as a pet, or for a 4-H project, but registration is a good way for the ILR to track the growth of the industry.

Blood-Typing

If you are buying a stud and plan to market him for outside breeding, be sure that he is blood-typed prior to purchase. Blood-typing must be understood to work on the principle of exclusion. Blood-typing will tell who WAS NOT the sire or dam of a cria in question. It will not prove that a specific sire or dam was in fact the parent, but it will indicate that the blood type of a particular cria (baby llama) is consistent with a possible progeny of those two animals. Blood-typing reveals any linebreeding or inbreeding, either deliberate or by accident, on the part of the breeder. The registry requires that any males used for outside breeding be blood-typed in order to register their offspring. (*Outside breeding* occurs when studs are used for hire for their reproduction services. Females can be shipped to the stud's farm for a set fee, but the stud can be taken to the female's farm as well.) Even in a hand-breeding operation (where the male and female are confined for the purpose of mating), it is good management practice for a breeder to blood-type *all* her animals. (See Chapter 9 for a discussion of breeding.)

Older Breeding Females

If you're shopping for an older breeding female, ask:

- ◆ Can you provide me with her reproductive history?
- ◆ Has she had any problem births?
- ◆ Is she a good mother, and does she provide plenty of milk?

When in the market for females, consider older breeding females. In many cases, these have been the foundation animals a breeder has utilized for many years in his breeding programs. (*Foundation* females are those that a particular breeder started with or first purchased as beginning stock.) They may be culls simply because they are getting up in years. (*Culls* are llamas that are no longer used for a breeding program.) If you are willing to take a chance, an older female can offer you a solid foundation on which to base your

Gale Birutta

This older female continues to produce champion offspring year after year.

breeding business. A female may have been culled because she tends to throw a color that is not particularly desirable to its owner. She could be an excellent breeding female, however, in *your* program.

Females can reproduce until about 25 years of age. There have been some reports of a female producing at age 35. If she is well cared for and not put under too much stress by an overly aggressive younger male, a female is seldom too old to breed. If she develops difficulties in birthing or other physical disabilities, such as back and leg problems, cease breeding her. This is an individual call with each female; some may develop problems at age 10, and others may produce until they pass away quietly in the pasture.

Pack Llamas

If you're shopping for a pack llama, ask:

- Is he pack trained?
- How much weight is he comfortable carrying?
- Is he a leader in the string?
- What type of terrain has he worked on?
- What distances is he used to?
- Is he trained to tether? (*Tethering* involves tying a llama to a ground stake with approximately 15 feet of lead for grazing.)
- What is his trail behavior? Does he work well with strangers? with other llamas?
- How recent is his conditioning?
- Has he been gelded? When?
- How easily can he be caught and haltered?
- Does he enjoy working?

Llamas selected as work animals also must meet certain criteria. A llama that is excessively large won't necessarily make a sound packer. A tall, lighter, well-conformed, and balanced llama may outperform a large llama. While geldings generally make the best packers, a manageable stud when taken away from females at the farm can also make a great packer, as can females who are not being used for breeding. There is no answer as to whether a stud, gelding, or female is the best packer, but generally geldings are quieter in nature and pay more attention to the job at hand than do studs. And because of the breeding value of females, they are not normally preferred for packing.

You have several options when choosing a pack llama: a trained and proven pack animal; a young, unproven animal; or an older, sound, but untrained llama.

A llama used as a pet or 4-H project should be well trained and like children.

Typically, low-end males fail as potential packers, and should be reserved strictly for the pet market. *Low-end* males are males with poor conformation, a bad attitude, or health or genetic problems. Typically, these llamas are inexpensive or given away for good reason. These animals are not often good packers.

Pets, Companions, and 4-H Projects

If you're shopping for a pet, companion, or 4-H project llama, ask:

- ◆ Has this llama been handled by children?
- ◆ Is this llama easy to catch, halter, and work around?
- ◆ Does this llama have any vices?

Llamas used for these purposes don't have to meet strict criteria. The most important factors here are personality and manageability. These llamas should be trained to stand easily for grooming and toenail clipping. They must be well halter- and lead-trained, and they must like children.

When considering a young llama, find out whether he is accustomed to handling by children. If a young llama is used to children, he will have a better attitude, not be spooked by quick movements, and be trustworthy with youngsters. Older or retired llamas can be used as pets so long as they have been handled. A manageable llama is one that is easily caught and haltered, and will tolerate grooming and fussing. Any llama you consider should be used to other pets, such as dogs and cats.

Guard Llamas

If you're shopping for a guard llama, ask:

- ◆ Has this llama been housed with other livestock?
- ◆ Is this llama gelded and over the age of 18 months?
- ◆ Is this llama aggressive toward members of the canine family?

You'll need to do some research to determine whether a guard llama will suit your needs. Most llamas will guard other livestock naturally, but with varying degrees of effectiveness. (For more on llamas as livestock guardians, see Chapter 15.)

Alpacas versus Llamas

Much of the information in this book can be applied to alpacas. General care requirements and management and breeding techniques are similar. Llamas and alpacas are similar in their worming and inoculation needs for the few diseases that camelids are susceptible to. However, llamas and alpacas come from different gene pools, so there are some important variations that must be addressed. Llamas and alpacas have different levels of athletic ability, fiber type, and dietary needs. Their responses to environmental and/or genetic problems vary.

One advantage of raising alpacas is that you can do it on a smaller scale. Alpacas require approximately ¼ less pasture than the same number of llamas and approximately ¼ less feed per head. One person can handle an alpaca more easily because they are smaller. In addition, if you are a sheep producer, few changes are required to house and fence alpacas.

Llama Purchase and Sales Agreement

Agreement of Purchase made this _____ day of _____ by and between the parties identified in Paragraph 1 below. PURCHASER agrees to buy and SELLER agrees to sell on the terms and conditions of this Agreement hereinafter set forth below:

1. **PARTIES:**

The party referred to in this Agreement as the "SELLER" is:

_____

The party referred to in this Agreement as the "PURCHASER" is:

_____

2. **DESCRIPTION OF LLAMA to be purchased:**

_____

_____

3. **PURCHASE PRICE:** _____

4. **PAYMENT & CONVEYANCE:**

   a. A deposit of _____ has been recognized and the remaining balance of _____ will be made at time of conveyance. SELLER agrees to deliver said male llama no later than _____ , at which time full balance is due.

   b. SELLER agrees to supply executed ILR certificate to PURCHASER at time of conveyance. SELLER shall supply halter and lead for said llama. Said llama will be show groomed for delivery.

5. **REPRESENTATIONS OF SELLER:**

   a. SELLER represents that she/he is the owner of said llama and represents she/he has not entered into any other agreement for sale of said llama.

   b. SELLER represents said llama is in excellent veterinary health and current on all inoculations and complete veterinary records shall be supplied to PURCHASER at time of conveyance. SELLER shall provide and absorb the cost of a veterinary health certificate required for out of state travel.

   c. SELLER provides with this sale a Warranty of Reproductive Capacity for said llama. SELLER guarantees that said male llama is at present reproductively sound and guarantee same to be a normal, reproducing intact male llama to be used as breeding stock. Should said llama be deemed to be incapable of reproduction by a licensed veterinarian, then SELLER will exchange the llama for a comparable intact male at no cost.

6. **ENTIRE AGREEMENT:**

   The parties agree and represent that this contract sets forth the entire agreement between the parties. It is further agreed that any modifications to this agreement be executed in writing as an addendum.

_____

DATED   Seller

_____

DATED   Purchaser

Seller's Responsibility

Customarily, it is the seller's responsibility to supply a health certificate to the buyer and to absorb its cost. A health certificate is merely an inspection form required by each state's Department of Agriculture. These are only required when transporting llamas across state lines. Each state has its own requirements regarding testing. Be sure to check with your veterinarian to learn about these tests and regulations. For informational purposes here, every llama is required to have a health inspection performed by a licensed veterinarian within 30 days of transport. Some tests that may be required are brucellosis, blue tongue, anaplasmosis, or tuberculosis (T.B.) Some states may also require proof of rabies vaccine or worming. But again, check with your veterinarian.

If you are transporting a llama in-state, you won't need a health certificate. A responsible breeder will, however, always supply a complete veterinary record that will include breeding records if applicable, along with worming and inoculation records. In some cases, a particular breeder may ask for a veterinary health certificate from a llama in-state. This is not required, but may be a precautionary measure on the part of that breeder.

Purchaser's Responsibility

As the purchaser, your responsibilities involve doing your homework. Be sure to get a written Purchase and Sales Agreement that spells out the terms of the sale, including a full description of the llama, the intended purpose of the animal, specific guarantees provided by the seller, and the conditions of sale. If the llama is a bred female, this should be guaranteed by supplying a progesterone test or an ultrasound result to the new owner. Should the female absorb or abort the fetus, a rebreeding should be supplied at no cost. A bred female is normally guaranteed to deliver a live cria. (A *live cria* is defined as a baby llama that has stood and nursed within the first 24 hours.)

With the sale of any breeding stock, the seller should guarantee that her animals are reproductively sound. For females, the agreement should be specific: Should a maiden female not be reproductively sound through no fault of the buyer, then an exchange for a like female should be guaranteed in writing. In any agreement, replacement of the animal or refund of money should be forthcoming if the animal is not as represented by the seller. As the buyer, it is your responsibility to see that the animal meets your criteria — it is the seller's responsibility if the seller *represents* that the animal meets the specific criteria when it does not.

Gale Birutta

Discriminating buyers look over a herd sire.

Pre-Purchase Exams

Although a veterinary pre-purchase exam is highly recommended, it is no guarantee that an animal will meet a buyer's particular needs. A pre-purchase exam will help ensure that the animal is healthy and could be used for its intended purpose. While responsibility for the veterinary exam is negotiable, it is normally handled by the seller. In cases where a considerable amount of money is being transferred, then the buyer may elect to pay for his own veterinarian to examine the animal. These exams can run anywhere from $50 to $150 or more. Some of the preliminaries start with you, the buyer. Ask the breeder to supply you with the veterinary records. Make sure all deworming and inoculations are up to date. (See Chapter 5 for a complete list of recommended vaccinations.) Inquire if the animal has suffered any injuries or illnesses that are not listed.

Examine the llama very closely: Is it developing appropriately for its age? Is the llama's head in proportion with its body? Are the eyes or nose running? Is the llama structurally correct? (See Chapter 5 for a discussion of conformation and soundness.) Ask if there have been any genetic defects in the bloodline, even if they are not apparent.

Body Structure by Age

Age	Male	Female
1–6 months	Short back Short cannon bones Slim barrel; able to feel ribs	Short back Short cannon bones Slim barrel; able to feel ribs
7 months–2 years	Back lengthens Cannon bones increase in length Barrel broadens and mid-section fills out	Back lengthens Cannon bones increase in length Barrel broadens and mid-section fills out
2–4 years	Neck thickens Leg bone density increases Overall body fills out	Neck thickens Leg bone density increases Overall body fills out
4–6 years	Male has reached full maturity (has stopped growing)	Female has reached full maturity Back may start to sway slightly from pregnancy
6–15 years	No change	More pronounced sway in back Belly and barrel size increase May develop dropped pasterns
15 years and up	Back may sway slightly Barrel thins out May develop dropped pasterns if has been heavily packed Lower front teeth may protrude	Sway in back becomes even more pronounced Belly size increases, especially if bred for many years Dropped pasterns become more evident Lower front teeth may protrude

Now contact your veterinarian. If the animal you're contemplating is local, use your own vet. If you are looking out of your own area, of course, this may be impossible. To avoid any conflict of interest, hire an independent veterinarian (one not affiliated with the seller's farm) who is familiar with llamas to inspect the animal you intend to purchase.

Tell the veterinarian that you're planning to purchase the animal and what its use will be. This will give him an indication of the type of exam to perform. For example, an extensive examination of the reproductive system should be done if you are purchasing a breeding animal. In a female, this exam includes a blood progesterone test or a rectal palpation to determine pregnancy. If you are considering a castrated male, then a less formal examination is required. To assist the veterinarian in evaluating the animal, supply any information you have obtained such as birthing problems in the female, illnesses or diseases, even broken limbs. Such health information should be supplied by the seller.

General Examination

The veterinarian should first perform a visual exam. He will be looking at conformation, structure, balance, and if the animal is the correct size and weight for its age. The veterinarian should also ask that the animal be walked in order to observe how the animal tracks (how and where the animal places its feet on the ground) and moves.

The next step will be close examination of the llama's legs, particularly if the llama is young. Although some immature llamas are knock-kneed (a condition that usually straightens out by maturity), certain abnormalities might be a sign of nutritional or genetic problems. Inspection may reveal arthritis or dropped pasterns (the part of the llama's foot between the ankle and the toes) in older llamas.

Heart problems are not unusual in young animals, and the veterinarian should listen to the llama's heart for abnormal sounds. (Heart murmurs are common in young camelids, but are quickly outgrown.) He should also listen to the lungs, to detect problems such as pneumonia. Nasal discharge could be

Some knock-kneed young llamas will grow out of this phase; others may be suffering from nutritional or genetic problems.

an indicator of several problems, such as respiratory infections, nasal bots (the larvae of the bot fly), or an allergic reaction to a particular forage.

The vet should inspect the llama's mouth to see if fighting teeth are still present. (*Fighting teeth* are two upper pairs and one lower pair of teeth in both males and females. They are sharper and pointed. To exert dominance, males use these teeth in competition.) For safety, these teeth should be removed at the jawline by a veterinarian. A llama's age can be determined by examining the teeth. As a llama gets older, the front, middle, and corner incisors lengthen. In some cases, these longer older teeth will have to be cut down so that the llama can eat more efficiently. A llama's age cannot be exactly determined by teeth, but within three years is a good bet. The vet should also inspect the animal's eyes for trauma, cataracts, or infections, and its ears will be examined for infections and abnormalities.

North American llamas are sometimes overweight (see Chapter 3). A veterinarian, running his hand down the neck and along the spine and ribs, can assess the animal's approximate weight. He should be able to feel the llama's spine — it should not be lined with fat deposits. He should be able to distinguish the ribs; the ribs cannot be found if there is considerable fat. Looking at the llama from the rear, the vet should be able to see the llama's belly, but it should not be bulging.

The veterinarian should also visually check for any external parasites. He will collect a fecal sample to check for the presence of internal parasites.

Male Reproductive Exam

A careful examination of a male purchased for breeding will help ensure that he can fulfill his role. Whether you are purchasing a young male as a future sire or as a proven sire, there are some factors that help determine breeding soundness.

It is difficult to note any breeding behavior in the weanling or yearling male. At this young age the *prepuce* (the sac that holds the penis) is still attached to the penis; thus, it is not possible to examine the penis. You can only examine the testicles by palpation (feeling with the hands). At birth, the testicles are in place in the scrotum, located under the anus. While the testicles are small, they can be palpated. If the testicles do not appear to be present in the scrotum, this could be because llamas have the ability to pull them forward when they are frightened or cold. If you can't feel the testicles, the llama may have a genetic defect. It is important to locate the testicles if you can. Reach forward of the scrotum from behind along the midline and slide your fingers back toward the scrotum while applying pressure. If testicles are present, they will ease back into the scrotum. At this time, examine them for uniformity of size and shape. A llama with no apparent testicles should never be purchased as a breeding male.

Beginners should rely on their veterinarian to examine the male's reproductive anatomy as part of the overall examination.

If you can observe a mature adult male during an actual breeding, you'll get a good idea of his behavior. Again, testicle presence, size, and shape are of utmost importance. A male with only one testicle has a genetic defect and should not be used for breeding. Males with large testicles may throw female offspring with large ovaries and male offspring with large testicles. (Both a female with large ovaries and a male with large testicles may have increased breeding stamina and fewer birthing or reproduction problems.) The penis should be examined for lesions or swelling.

A breeding male may have 10 offspring on the ground, but this does not mean that he is still fertile. A male can become infertile for a variety of reasons: because of extreme heat (a temporary problem); an infection that may have developed; or because the male may simply be too old. When a considerable investment is involved, have your vet collect a semen sample for evaluation of sperm size and shape, concentration, and viability. Unusually small or irregularly shaped sperm may indicate decreased male fertility. Therefore, the buyer and seller should negotiate the responsibility for payment for such an analysis.

Kevin Kennefick

Normal male testicles are of uniform size and shape.

Female Reproductive Exam

Rectal palpation for reproductive capacity in the female llama is recommended for those over 18 months of age. Even though a female has already produced offspring, she may have damaged her reproductive organs in the most recent birth, or she may develop reproduction problems in the future. The veterinarian will be able to assess the overall condition of the uterus and cervix.

Palpation will allow the veterinarian to determine if the reproductive tract is intact. If the llama is more than 30–45 days pregnant, the vet will be able to determine that, too. (Pregnancy can also be confirmed through a simple blood progesterone test.) During palpation, in most instances, the left horn of the uterus will feel larger than the right horn if the female has previously given birth. The ovaries can be palpated for shape, size, and consistency, as well as to detect any uterine infections. The size and shape of ovaries are important. Females with larger ovaries have fewer reproduction problems. An ovary with an odd shape may indicate a follicular cyst, which can cause infertility.

Basic First Aid Kit

A simple first aid kit should consist of a 7 percent iodine solution for possible wounds, cotton, gauze wrap, disposable rubber gloves, and a digital thermometer. A household first aid kit can be adapted for barn use. All medications should be administered by your veterinarian until you learn how to do this on your own.

Anemia

Should your veterinarian suspect anemia, a complete blood count (CBC) should be done. The CBC includes a count of the packed cell volume (PCV) and the number of red blood cells (RBC). Low PCV and RBC often indicate anemia. An anemic llama will require additional care to recover. Anemia could be a sign of poor farm care and management. Although anemia isn't a fatal condition and an anemic llama can regain its health, it is best to look around for a healthy llama. But if this is the llama you really want, let the seller take care of the condition. You can come back and reexamine the llama at a later date.

An abnormally low or high white blood cell (WBC) count may indicate an infection. The white blood count can tell you if there is a severe parasitic condition. Don't consider a llama with parasites. While some parasites are species specific, others are not, and you don't want this llama to infect other animals in your stock. Be sure any llama that has been diagnosed with any type of parasite has a clean bill of health before leaving the host farm.

Veterinary Certificates

Upon completion of the final phase of the examination, the vet will issue a certificate showing what tests have been done on that animal. If the llama is from out of state, the veterinarian will take care of all tests and the paperwork associated with transporting the llama to its final destination. Be sure to notify your veterinarian at least three weeks before your departure; this way, there will be no last-minute surprises or delays. Follow the requirements of the state to which the llama is going.

Preparing for the New Arrival

Making basic preparations at home need not be complicated. If you are bringing your new llama home during the spring or summer months, any simple lean-to is sufficient. Colder, windier climates require a more substantial three-sided shelter. A formal barn is not necessary for llamas, but many serious breeders will eventually opt for one.

Fencing need not be elaborate; it can even be temporary if you are anxious to bring your new llama home sooner. Portable sheep fencing can be purchased at most agricultural supply stores and installed easily. If you are planning to use existing fencing, make sure it is in good repair. Walk your pasture, pen, or field and fill in any holes to prevent injuries.

Your new llama should always be supplied with a clean and dry place to lie down. Straw or low-quality hay, like mulch hay, is perfect for bedding. A horse trough, or even just a plastic or rubber bucket, is fine to hold drinking water. Plastic feed pans can be purchased at feed stores or at llama supply houses for about $3 each.

The seller should provide you with a feeding schedule and note the type of feed that your new llama has been used to eating. Many breeders will send along a 50-pound bag of feed, along with a couple of bales of hay. Changing grain is usually necessary, but most major feed manufacturers offer compatible llama pellets. Feeding time should reflect the llama's previous schedule for a few days, but can then be gradually changed to accommodate your schedule. Hay consistency is not generally a problem either, unless the llama is being moved to a different region. But llamas adapt quickly to such a change.

Basic Barn Equipment

A rake for manure
A shovel
A broom
Toenail clippers
A hoof pick
Brushes for grooming

Transporting Llamas

Llamas are good travelers and can be transported in a variety of vehicles. Depending on how many llamas you are moving and their sex and age, some may cohabitate well on the journey. Keep open females and studs separate, though, and dams and their crias should travel only with other dams and crias. This way the babies won't be injured by the adult males.

You can easily train llamas for loading. Generally they will *kush* (lie down) while on the road. Llamas usually travel in horse trailers, vans, and pickups. (A tractor trailer driver passing through Vermont is still talking about the llama in the Suzuki Samurai!) Pickups and vans may have slippery floors and will need a rubber mat or carpeting to avoid injury to the llama upon loading and unloading. If you plan to transport llamas in the back of a pickup truck, make sure the siderails are high enough to prevent the llama from jumping over. Any transport vehicle should have ample bedding, water, and hay for the trip.

Usually, you won't need to tie llamas — they like to move around. If they're traveling with a buddy, they'll want to be close to their friend.

To train a llama to load into a vehicle, start when the llama is young. Lift him into the vehicle occasionally until he becomes accustomed to it. Later, lead the llama to the vehicle and allow him to

Gale Birutta

A pack llama eagerly jumps into the back of a 1960 International Metro Van.

explore. Be very patient as you encourage the llama to go a little bit further, then a little bit further into the vehicle. If possible, have someone assist you by standing behind the llama with arms outstretched to prevent the llama's moving backward and away from the lesson you had planned. (A young llama will learn readily by watching an experienced llama being loaded.)

After he has gone into the vehicle, close the door (or the gate) and praise the llama. Allow him to spend 15 minutes getting used to being in the vehicle. Then turn him around and open the door (or gate) so he can jump out.

Repeat the lesson several times a day for one week. Then take short rides to acclimate him to travel.

Innovative Llama Loading

Moving day was here. Our personal belongings were no problem, but our female llama, Jamaika, was due to deliver in less than two weeks. This would be only the second time we had trailered her, and we knew we were in for a battle. First we tried to get her into the horse trailer by herself. No go. Perhaps if we put Monte in first, she would follow. Of course not. She was bred and she wanted no part of sharing a trailer with the male. After about an hour of coaxing, John decided to get one of the large moving straps (the straps that are about 3 inches wide). We slung it behind her hind end, attached it to the inside front of the trailer, and winched her in. Innovation goes a long way when you need it to.

Vans

A full-sized, extended cargo van can carry as many as four adult llamas, and a minivan with its rear seat removed can accommodate two adults. Vans provide an excellent opportunity to monitor your llamas' behavior and comfort while traveling. Keep llamas separate from the front seat with a partition. Llamas love to see where they're going and will find their way up to the front, presenting a driving hazard.

Pickups

Pickups are high off the ground. Backing up to an embankment or grade to help coax in a llama that hasn't previously been trained can help. Stock racks that are 3 to 4 feet high are inexpensive and easy to install, but slats should be close enough (3 to 6 inches) so the llama's neck and legs won't get caught. For long-distance traveling in foul weather, a cover is recommended.

You should avoid transporting llamas over long distances in open pickup trucks in cold climates; the wind chill will make conditions too uncomfortable for your animals.

Horse Trailers

Horse trailers are low to the ground for easy boarding, and most are equipped with a wood or rubber floor. There's usually room for as many as four llamas. Take care when transporting llamas in a trailer with a center divider; llamas have been known to climb over the divider and become injured. You can leave llamas in ventilated horse trailers for several days when you are transporting them on a long trip. It is essential to provide fresh feed and water twice a day. You won't have to remove llamas from the trailer to walk them. These are quiet, contented animals that have no problem staying in a kushed position hour upon hour chewing their cud.

Station Wagons

Don't forget the convenient, all-purpose station wagon. Nothing beats it for single-llama, short-distance transport. If a llama has traveled in a station wagon in his younger days, he will continue to "fold up" nicely as he grows older. (Of course, this mode of transportation won't be comfortable for very large llamas.)

Bringing Your Llama Home

You have chosen your first llama, and now it's time to bring him home. You should sit down with the seller and review with her what you will need in the first few days.

Your new llama should come with a properly fitting halter and lead. Some first-class breeders will show groom your new llama so that he looks his best. Toenails should be recently trimmed. Having these chores taken care of by the seller will alleviate some of the stress you may feel, and you'll appreciate the effort: You don't want to start out with a matted llama with overgrown toenails.

A sympathetic seller will supply you with several days' feed to allow you to gradually change him over to a new diet if needed. Check with his previous owner on feeding practices. The seller will tell you the type and amount of feed the llama is used to.

Home at Last!

Day one: When your new llama arrives, take him for a short walk to stretch his legs. If you have a dog, be sure it is not around. (Allow dog and llama some time to get used to each other later.) Put the llama in his containment area and allow him to settle in for a day or two so he can become familiar with his surroundings. If other llamas or livestock are present, keep him close to them but not in with them the first day. Let him adjust. Provide fresh water and pasture (hay if it's non-foraging season). Try not to change his feed, unless your llama is now in a different part of the region or country. If he's used to grain, you may have to change his diet to suit your budget or schedule.

Day two: Chances are your llama has adjusted already. (Some may require an additional day or two.) Introduce your new llama to other llamas or livestock over the fence. Allow them to sniff each other. Take your llama for a walk around the farm to familiarize him with more of his new surroundings. Again, if you own dogs, be sure to keep them out of the llama's way. If your llama is used to dogs, you can have someone walk the dog on a leash nearby until your llama accepts the dog.

Insurance

Many insurance companies will not cover losses that involve livestock. For instance, homeowner's insurance will not cover barns or other structures against fire if hay is stored in them. As soon as hay is added, you are considered a farm and your homeowner's policy should be switched to minifarm or farm insurance. Unfortunately, even many *experienced* breeders are not aware of this. Contact your insurance company and find out what is and what is not covered under your current policy. You will probably need to change your insurance company to one that covers livestock and farms.

A standard farm insurance policy will cover certain named perils to livestock. These include drowning, lightning, predator attack, accidental shooting, electrocution, and loading and unloading accidents. Death from disease is not covered.

You need to itemize the value of each animal. Because of the investment for females and in some cases a valuable breeding male, you may want to secure additional mortality insurance. This way, if your llama dies of any cause, you will recoup your investment.

Day three: Continue walking your llama and showing him around. In most cases, he may mingle a bit with other llamas. As he adjusts, you can start working with him and handling him on a more regular basis. Within a week, he should be fully settled in and part of the herd.

From now on you should work with your llama daily. Catching, haltering, and taking short walks will always benefit both of you. Daily grooming helps to desensitize your llama. (*Desensitizing* means gently touching your llama all over his body: face, head, back, legs, and belly.) A llama will readily accept a soft brush around the face before he'll accept hands. Most llamas are not as well desensitized on their legs as they should be. Again, regular, gentle brushing on the legs will relax your llama and soon you'll be able to work around his legs.

THREE

NUTRITION AND FEEDING

Llamas are *modified ruminants*; that is, they chew their cuds but do not have rumens (the first stomach compartment, in which cellulose is broken down) — they have only a three-chambered stomach. There are three essential elements in a llama's diet: protein, vitamins, and minerals. The animal obtains these elements through forage, supplements, grains, and mash (for older llamas that may have lost teeth).

Proper nutrition is important for all animals at all stages of development. Nutritional needs vary for young animals, pregnant and/or lactating females, breeding studs, working llamas, and inactive animals. Be aware of the nutritional content of your forage; your llamas may need a mineral supplement. Any agricultural lab or extension service can perform an analysis of forage. Soil content, the time of year for pasture (stage of growth), and how long hay has been in storage will affect the nutritional makeup of your forage.

A variety of forages, properly managed, will supply most of a llama's nutritional needs. If supplements are needed, provide grain and free-choice salt and minerals.

Protein

Because llamas are ruminants, their protein requirements are lower than most other species. This is due to the above-average efficiency of their digestive system in converting lower feeds and forages into protein. Protein needs do, however, vary. Mature llamas on an average maintenance diet require 8 to 10 percent protein contained in their feed. Ten to 14 percent protein is recommended for pregnant and lactating females. Growing babies will need 10 to 16 percent. Keep in mind that overconsumption of protein is detrimental to a llama's health. When too much protein converts to fat, it results in obesity.

Overweight females can develop delivery difficulties; overweight breeding males may have trouble breeding and/or be more susceptible to heat stress. The following chart summarizes llamas' protein needs throughout their lives.

Protein Needs of Llamas

Age/Condition	Percentage of Protein in Diet	Source
Crias up to 6 months	10–16*	Colostrum; legumes
6–8 months	10–16* (16% recommended)	Legumes; llama feeds; cottonseed meal
24–48 months (male or open female)	8–10	Grass hays; llama feeds
Pregnant females**	10–14	Llama feeds; green grass; legumes; cottonseed meal
Lactating females (birth through weaning)	10–14	Llama feeds; legumes; green grass
Over 48 months (male or open female)	8–10	Llama feeds; grass hays
Pack llamas (worked 4–6 hours, 3 times per week)	10–12	Grass hays; llama feeds
Guardians	8–10	Green grass; llama feeds
Breeding studs	10–12	Llama feeds; grass hays; green grass
Inactive or elderly llamas	8–10	Llama feeds; grass hays

*Avoid overfeeding of protein; it converts to fat, which is detrimental to the llama's health.
**In the third trimester, a minimum of 14% is required.

Vitamins

Vitamins are also essential to a llama's diet; in particular, vitamins A, B-complex, C, D, and E. Without proper amounts of these vitamins, serious health problems can result. The following chart summarizes the role each vitamin plays in the llama's physical health.

Essential Vitamins in Llama Diet

Vitamin	Importance	Signs of Inadequate Intake	Source
A	Growth and bone development	Fiber and skin problems	Colostrum*
	Proper tooth formation	Sterility/infertility	Green pasture
	Lactation	Inadequate milk supply	High-quality hay
	Fertility	Angular limb deformity	Llama feeds & grains
B-complex	Maintenance of healthy nervous system	Uncontrollable shaking or twitching	Llama feeds & grains; legumes
	Formation of red blood cells	Anemia	Healthy llamas produce in digestive tract
C	Tissue formation	Weak muscles	Legumes; carrots; vegetables
	Proper bone growth and tooth formation	Poor teeth	Healthy llamas produce in liver
D	Bone growth	Angular limb deformity	Llama feeds
	Proper tooth formation	Poorly formed and missing teeth	
	Calcium absorption		
E	Fertility	Sterility/infertility	Llama feeds & grains
	Maintenance of normal red blood cells	Anemia; skin problems	

*Colostrum (the dam's first milk) is an essential source of Vitamin A for newborn crias, as well as containing disease-fighting antibodies for the new babies.

Keep in mind that you may have to provide vitamin supplements. For example, while a healthy llama should produce vitamin C in its liver and the B-complex vitamins in its digestive tract, deficiencies can still arise and need to be addressed. Taking responsibility for all of your animals' dietary requirements is of the utmost importance.

Minerals

There are five major minerals required in the llama's diet: phosphorus, calcium, selenium, copper, and zinc. The following chart summarizes their importance.

Essential Minerals in Llama Diet

Mineral	Importance	Signs of Inadequate Intake	Source
Phosphorus	Fertility Bone growth	Infertility / sterility	Timothy hay Clover/grass Orchard grass Llama feeds Cereal grains Trace minerals
Calcium	Bone growth	Poor or inadequate bone development Stunted growth Bone deformities	Llama feeds Oral supplements Trace minerals Legumes
Selenium	Fertility Healthy muscles	Infertility / sterility Paralysis, muscle stiffness, lack of coordination, difficulty standing, difficulty nursing	Oral supplements Llama feeds Trace minerals
Copper	Fertility Bone growth	Infertility/sterility Anemia Poor growth rate Skin problems Low-quality fleece	Oral supplements Trace minerals
Zinc	Fertility Bone growth and development	Infertility/sterility Poor bone growth Cracked foot pads Poor toenail growth	Oral supplements Trace minerals Llama feeds

Male fertility problems — decreased libido; delayed puberty; poor sperm production
Female fertility problems — failure to conceive; abortion; stillbirth; retained placenta

Other minerals that are important in a llama's diet are magnesium, manganese, and iodine. Magnesium and manganese affect bone growth and development, while manganese and iodine contribute to fertility. Although these minerals are not as crucial as the above-mentioned five, by feeding your llamas trace minerals, oral supplements, or commercial llama feeds, you will be certain that they are receiving what they need.

Salt and minerals are available in a free-choice mineral mix. However, be sure your area is not overly high in any mineral before giving supplements. Excess selenium, in particular, can be toxic to grazing animals. But too much of any mineral can cause health problems for your llama. Have your soil tested first, then assess the need for supplements based on the results.

Grain

Whether to feed or not to feed grain should depend on the quality and quantity of available forage. Most of the major grain manufacturers supply grain that is *suitable* for llamas; however, grains *specifically manufactured* for llamas are best. Agway, Blue Seal, and Purina Mills are among the leaders in the production of llama feed.

Kevin Kennefick

Mixtures of grains, such as sweet feed/corn/rolled oats/bran, are nutritionally sound. You can find them in the form of llama pellets, too. Cut apples, carrots, and other vegetables are also tasty and nutritionally beneficial. Most animals relish some variety in their diet. Supply grain mixtures and fruit as part of the regular diet in the form of a handful a day or as a treat when working or training your llamas.

Feeder Designs

Indoor hay feeders: Hay feeders inside barns and shelters are relatively easy to construct. These feeders can be built against an entire wall of a stall or holding area, using the corners for the main support. Height from the ground should be between 3½ feet

Llamas love to browse on brush.

and 4 feet. By utilizing 1 inch x 2 inch boards spaced 2 to 4 inches apart and nailed to a 2 x 4 nailer, this slatted feeder will accommodate both adult and younger llamas.

Outdoor hay feeders: The same basic design for the inside hay feeder can be easily adapted to meet outside needs. With "V" shaped design, utilizing the same end bracing with 2 x 4's and 1 x 2's spaced 2 to 4 inches apart, you can make a feeder that can be used outside and carried inside (if space permits) for indoor feeding as well. Because of the design of this feeder, llamas can utilize both sides. This feeder can be modified to extend anywhere from 5 to 8 feet in length, and to accommodate up to 8 llamas at one time. If the feeder is left outside, be sure to cover the plywood roof with tar paper or some other roofing material. This will lengthen the life of your feeder. A tray at the bottom of the "V" configuration will catch hay droppings for the llamas to clean up. Without a tray, hay will fall to the ground and be wasted.

Grain feeders: Grain feeders can be simple to obtain, such as plastic pan feeders. They are available from any llama, agricultural, or feed supplier and cost about $3 each. These pans permit you to feed your llamas anywhere, without having to rely on permanent feed boxes. They are easy to clean as well.

Horse feeders are more permanent; these can be either plastic or rubber, and can be utilized on the ground or installed in the corner of a stall. These are also available at any feed store.

Wooden feed boxes may be made from one-inch lumber — softwood lumber is the easiest to work with. The ends and joints can be first glued together with a wood glue, then nailed with finishing nails. Be sure to countersink all nailheads to avoid injury. Wood screws may also be used, but again, be sure to countersink them. These can be fastened directly to shed, shelter, and barn walls.

Mineral or supplement feeders: Plastic or rubber feeders can be utilized for trace minerals or supplements. These feeders must always be used indoors as the minerals and supplements need to stay dry. Commercially made mineral feeders that can be left outside are available from farm supply stores. Keep in mind the ages of your llamas, and remember that several feeders should be placed between 3 and 4 feet high to accommodate the younger ones, too.

Hay

Choosing and feeding the proper hay will contribute to your llamas' health and nutritional balance. The basic considerations are your herd's overall health and the least waste at feeding time. Other factors will also contribute to choosing the proper hay quality for your herd: exercise and work, herd management, and region of the country. Quality of hay directly affects its value as feed.

Hay quality is affected by the stage of maturity at which it is harvested, its leafiness, odor, and color. The amount of foreign matter such as weeds, sticks, and other undesirable materials also affects the overall value. Quality hay possesses high amounts of protein and total digestible nutrients (TDN). Llamas require only about 10 percent protein for a maintenance level, but the TDN should be 55 percent.

Comparison of Hay Crops

Crop	% TDN	% Fiber	% Protein	Comments
Alfalfa*	55–60	23–29	15–18	Too much daily protein converts to fat, which causes obesity. Not recommended for llamas.
Brome grass*	59–68	30–37	10–16	Good choice; provides proper balance of TDN and fiber, calcium/phosphorus.
Clover/grass**	55	29	16	Excellent source of phosphorus; use as supplement only — will cause obesity; too much protein.
Oat hay*	53–72	27–29	11–18	Highest in TDN; best choice for cold-weather feeding.
Orchard grass*	54–65	31–37	8–15	Good choice; provides proper balance of TDN and fiber; calcium/phosphorus.
Timothy*	54–58	31–33	8–10	Good choice; provides proper balance of TDN and fiber; calcium/phosphorus.

* Percentages vary with season and region.
** Percentages are averages.

With legumes (alfalfa and clover), the leaves may fall off in later cuttings. Fully two-thirds of the protein of legumes is found in the leaves, and when it is harvested at or after maturity, the stems enlarge and the leafiness of the plant declines. When you are choosing legumes, it is a good idea to look for leafier hay, smaller stems, and bright green color. Alfalfa generally has approximately 16 percent protein. Although clover is usually mixed with some other type of forage, it carries roughly 16 percent protein as well.

It is important to harvest grass hay before full bloom so there are no ripe, mature seeds. When it is cut too late, grass hay stems will be coarse and yellow rather than green. Be aware that brome grass hay, which matures more slowly than most other types of grass hays and thus keeps yield high, has a fairly high fiber percentage of up to 37 percent, 68 percent of TDN, and 16 percent protein. This is an excellent mixture for llamas.

Color is not always an important factor. The color of good quality hay may simply fade from the sun. Sun-bleaching only occurs on the outer edges, however. On the other hand, wet hay will be brown or black, and should be avoided. You should be careful to steer clear of overly mature hay as well.

Use high protein hay sparingly with llamas; it can cause obesity and kidney problems. High-fiber hays such as oat hay are preferable. First cuttings of hay are generally well suited for llamas — not overly high in protein but supplying enough fiber. Llamas will readily eat second cuttings of hay, which are much higher in quality and higher in protein, but these can cause them to put on too much weight.

Proper hay storage is simple but important. Make sure the hay is dry before storing. Hay should be kept up off the ground, preferably in a loft. Check for leaking roofs and broken windows that may allow moisture to enter. Excessively green hay that has not been properly dried by your hay supplier is highly dangerous. Spontaneous combustion can result in fire. Obtaining hay through a reputable and experienced hay dealer will reduce or eliminate any problems. If the dryness of hay is in question, store hay on its side to allow any moisture to escape.

Excessive light can rob hay of its nutrients, so hay should be kept in a dark area. Again, the typical hay loft is ideal for storage. Hay also tends to lose its quality after a year. However, second cutting hay that is a year old and has been properly stored is sufficient for llamas' unique digestive systems. Llamas are one of the few livestock species that can utilize hay of this type.

Feeding Practices

Under some circumstances you may have to change the type and amount of feed for your llama. Pregnancy, illness, and the introduction of llamas from another part of the country all dictate the use of special feed. Make changes gradually. Base feeding decisions on the quality of forage and the condition of the animal.

Mature Llamas

Mature llamas (over 2 years old) usually range in weight from 250 to 400 pounds; a few may weigh 475 pounds or more. Daily forage should be equal to 2 percent of a llama's body weight. Working llamas and breeding males may require additional amounts. Older llamas that have lost some of their teeth should be fed only second cuttings of hay. First cuttings are too coarse and stemmy for them to chew properly, making digestion difficult.

Older Llamas

Older llamas are 12 years old and up. They may have lost some teeth, and therefore have difficulty chewing their food. Without proper chewing and digestion, older llamas may struggle to maintain their weight. Mash is a good solution.

Making a Mash

You can make a mash from a combination of horse sweet feed, llama pellets, and water. The proportions are ½ cup of warm water to approximately ½ to 1 pound of feed. Add ½ cup soybean meal and ½ cup cracked corn. Mix thoroughly and let soak for several minutes, then mix again. Feed twice daily.

Pregnant Females

Gestation for llamas is about 11½ months. Feed pregnant females in the same way as you would other mature llamas. It is especially important not to overfeed the female during the first two-thirds of her pregnancy. An obese female can experience delivery problems and decreased milk production.

The last trimester — the last 3½ months of pregnancy — is the period of most of the fetal growth, so an increase in nutrients is necessary. By increasing the ration of quality hay or forage and providing access to free-choice minerals, you will satisfy most of the llama's nutritional needs. If only lesser quality forage is available, supplement her feed with ½ to 2 pounds of grain per day.

Lactating Females

Lactation takes its toll. The female needs increased nutrients for milk production. By following the feeding requirements for late gestation, you ensure that she will maintain the proper nutrient levels. Decrease feed to a maintenance level after three months of lactation to prevent her from gaining too much weight. At weaning, when the baby is 4 to 6 months old, eliminate the grain and decrease the amount of hay. Milk production will decrease and nutrients will be diverted to her new pregnancy if she has been rebred.

Underweight Llamas

Underweight llamas are rare in North America, but on occasion — because of cold climates, lactation, illness, parasites, stress, excessive breeding, old age, or poor forage — it does happen. Supplement the diet with whole cottonseed, as cottonseed is high in protein and has a 20 percent fat content. Mix 1 part whole cottonseed with 1 part corn and 1 part alfalfa pellets. With its high fiber content, this ration will return the animal to its proper nutritional balance.

It's fairly easy to determine if a llama is underweight. A bony breastbone is typical of an underweight llama, as are protruding hip bones. Some thin llamas may also be anemic or have a parasite problem, but the majority of underweight llamas are just not getting enough to eat or the proper nutrition balance. Younger and less aggressive llamas may suffer from poor nutrition when they are at the bottom of the pecking order. Separate these llamas and feed them where they won't be intimidated or pushed around.

Obese Llamas

Obesity in llamas is the result of overfeeding of both protein and energy-producing feeds. Obese llamas are more susceptible to heat stress and may develop heart and circulation problems.

To reduce weight, curtail the feeding of grain and discontinue free-choice feeding of hay and/or pasture. It is also helpful to feed obese

Acceptable Weight Ranges

Age	Type	Weight Range
Newborn cria (crias should gain 1 pound per day first 2 weeks)	Male/female	16–35 pounds
Crias: 4–6 months	Male/female	75–125 pounds
Weanlings: 6–12 months	Male/female	125–175 pounds
12–24 months	Male/female	175–250 pounds
24–36 months	Open female/male	250–400 pounds
36 months and over	Pregnant females	300–450 pounds
36 months and over	Males (working or nonworking)	325–475 pounds

llamas separately to restrict and control intake. Supply only oat hay, with no seeds, at 1 percent of your llama's body weight. (Invest in a good scale. Llamas can be deceiving with their weight, especially heavily fleeced animals.) Increase exercise to promote better circulation and fat-reducing capabilities. Be careful not to overexert your llama; start an exercise program gradually. Because you are limiting forage and grain, be sure to offer free-choice minerals to ensure that your llama receives adequate nutrients. Rebreed dams as soon as possible after they give birth, and allow them to wean their crias later than usual.

Body Scoring

Perform a hands-on evaluation of body condition routinely. This technique will help you to evaluate whether your llama is the proper weight. The procedure for body scoring is as follows:

1. Feel down the backbone of your llama, starting at the withers. (The *withers* are where the shoulders meet the backbone at the base of the neck.) If your llama feels like a firm and padded mattress and you are unable to feel any backbone, your llama is too fat. If you can feel *each* vertebra of the back, your llama is underweight. Prominent pelvic bones do not necessarily constitute low weight; overweight llamas can have a bony pelvic area.

This llama is normal weight.

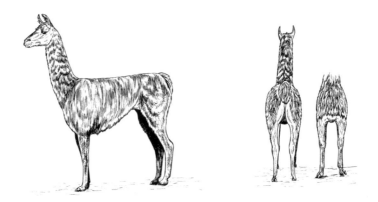

This llama is too thin.

This llama is overweight.

2. It is normal to be able to feel the ribs on a llama. Your llama is overweight if you cannot feel the ribs.

3. Look at your llama from the rear. You should be able to define the muscle mass in the thighs without seeing it shaking as your llama walks.

4. Feel for the breastbone; it should not be overly bony, nor should it have the well-padded-mattress feel of the back.

Cold-Weather Feeding

Deviation from the regular feeding practices during snaps of extreme cold is mandatory in some sections of the North. Temperatures of 15°F and below for more than a day warrant extra attention. Most llama breeders will increase feed and hay to ward off the stress of cold winters. Increasing general intake above the 2 percent level is pointless: The *type* of intake must be defined, not the *amount*.

Energy and Fiber Levels

The need for increased energy during bitter cold can be satisfied by adding total digestible nutrients (TDN). Llamas normally require a feed level of 55 percent of digestible nutrients daily for proper maintenance. In winter, that level jumps to at least 65 percent. High-quality hay or alfalfa will not meet the energy requirements during bitter cold. Energy supplements are available in commercial feeds and forage, in the form of carbohydrates, proteins, and fat. While proteins will create energy, some forages such as alfalfa decrease in fiber content as the protein level rises. The decreased fiber intake may result in digestive problems.

Fiber is an integral part of a llama's diet. Without sufficient fiber, the performance of the digestive system is challenged. A diet of approximately 25 percent fiber is recommended. Fiber reduction results when both protein and TDN are increased (as with high-quality hay). High-quality hays, such as alfalfa and second cuttings, contain high levels of protein because the hay is not allowed to mature enough to create heavier and coarser stalks with sufficient fiber.

Balancing TDN and Fiber

There are several ways to provide good balance in the cold-weather diet. Oat hay is an excellent choice to provide the proper amounts of fiber and TDN, but it can be difficult to find in some areas of the country. Generally

Cold-Weather Requirements

	Needed	Source
Fiber	25%	Llamas normally need no supplement other than mid-quality hay. If needed: add cottonseed; beet pulp; citrus pulp; crimped oats.
TDN	65% (10% more than usual)	Oat hay; brome grass hay; orchard grass hay.
Energy	No studies have yet been done to determine what percentage of overall body metabolism is needed for energy.	Flaked or rolled corn; 50% soybean meal and 50% corn as top dressing; commercially produced llama feeds.

Important: Always have fresh water readily available!

It is important to learn "body scoring" to determine whether your llama is maintaining its proper weight. Performed on a weekly basis, particularly during cold snaps, this procedure will also determine if the llama is cold and needs to generate more energy.

grown as a feed crop for alfalfa, it is harvested while still green (not allowed to dry) but almost at maturity.

Corn, either flaked or rolled, is an extremely high energy source that llamas enjoy. Although it delivers only about 9 percent protein, it will raise the TDN to 85 percent or even higher — a good supplement during periods of extreme cold.

Cottonseed possesses an excellent fiber content — 20 percent. While high in protein at 20 percent, the fat content of 20 percent will deliver the high energy llamas need.

Problems Relating to Poor Nutrition

Improper nutrition can contribute to a variety of problems. Don't confuse physical defects resulting from deficiencies in protein and energy in the diet with genetic problems. Analysis of your herd's nutrition may tell you that problems are nutrition-related. Sometimes physical defects in newborns and young llamas show up although the herd has not shown any before. This, too, could be nutrition-related.

Angular Limb Deformity

Nutrients involved with bone growth are phosphorus, calcium, zinc, magnesium, copper, protein, and vitamins A and D. Poor or crooked limb structure does not always constitute a nutritional problem, but clearly defined deformed bone structure is a matter that needs to be nutritionally addressed.

Vitamins A and D and protein can be supplied through high-quality forages such as alfalfa and clover. Pasture that is highly fertilized with potassium and nitrogen will most likely be deficient in magnesium, although these pastures may grow rapidly lush. Grains manufactured specifically for llamas will contain the proper amounts of magnesium, copper, and zinc.

Angular limb deformity (crooked legs) is common in baby llamas, most often caused by the looseness of ligaments on either side of the knees that help hold the knee joints together. If it is not genetic, time will usually correct this situation. If the condition hasn't corrected itself within the first several weeks, then it is time to pay more serious attention. Legs may also appear normally straight at birth, but deformities can show up later at 2 to 6 months. An x-ray is the only way to determine the cause of the problem.

Angular limb deformity is common in baby llamas.

Feeding/Nutrition Differences: Llamas and Alpacas

The alpaca's evolution in extremely high altitudes has provided him with the ability to digest green grasses rapidly. Therefore, the alpaca's digestive system can process forage with a higher water content faster than can the llama's system. Alpacas are able to secure their nutrients sooner during grazing.

Alpacas are generally more prone to phosphorous deficiencies than are llamas. Some recently weaned alpaca babies experienced walking problems and almost ceased to grow. Angular limb deformities became apparent. Although they are still not completely understood, these problems can be solved by giving a mineral and/or a feeding supplement at weaning. It is important, however, first to measure the phosphorous readings through a blood test to make sure that low levels are causing the problem.

Fiber and Skin Problems

Healthy skins and coats require basically the same nutrition as bones. Vitamin A, important for fiber growth and skin health, is supplied by quality green forage. Llamas lacking vitamin A will display erratic or excessive shedding; extended periods of copper deficiency will produce a fiber of lower quality with poor pigment. Low levels of vitamin A and/or copper may also be the problem if your llamas are experiencing poor toenail growth and/or cracked pads, as they are part of the skin as well. Zinc deficiency can also cause or contribute to skin problems.

Infertility

Proper levels of vitamin A, phosphorus, vitamin E, selenium, zinc, copper, manganese, and iodine are all necessary for fertility. Provide food that has the proper balance of protein and energy as well. Deficiencies in any of these may result in decreased libido, poor sperm production or delayed puberty in the male, and abortion, failure to conceive, stillbirth, and retained placenta in the female.

Pasture Management

To utilize grazing and forage areas most effectively, breeders and owners must assess their situation and allow for expansion in the future for proper pasture management. As many as five llamas can be grazed on an acre of pasture;

however, all their nutritional needs may not be met. Supplements would then be in order, but this is expensive. Through proper pasture management and grazing techniques, you can keep expenses at a minimum and your pastures will sustain healthier forage.

Assess Your Fields

To provide an accurate assessment of your grazing areas, walk your fields and examine the forage carefully. Bare spots are a problem — they are not producing forage and are an invitation to weeds. If the ground is covered with weeds, you'll need to improve the forage quality and quantity. Collect soil samples and have your extension service analyze them for any deficiencies. You'll improve the forage dramatically simply by liming in the fall and fertilizing in the spring. Lime sweetens the soil and fertilizer promotes plant growth. Allow at least one rain on the fertilized pasture before permitting your animals to graze there. If your pasture has dandelions and immature berry bushes, these favorites of llamas will supply good nutrition when grazed in the immature stage.

Gale Birutta

A healthy pasture should be grazed early in the season or mowed.

Undergrazing/Overgrazing

Undergrazing and overgrazing are common problems. You can improve the quality of your pastures by implementing a grazing management program.

Rotational grazing involves breaking the grazing areas into separate, smaller areas and rotating more densely stocked animals to utilize forage more efficiently. This can be done with "sheep netting," which is a type of rolled portable plastic fencing that is widely available in farm and feed stores.

Gale Birutta

The forage in this pasture is past its peak for grazing and should be hayed.

Pastures of ¼ acre or less can support five or six llamas for two days. In spring and early summer — periods of heavy growth — you may rotate llamas back on the grazed pasture within two weeks. As the summer lengthens and the growing season shortens, allow an additional week per month. For instance, if in June it took two weeks for regrowth (regrowth is defined as 2½ inches), you would be able to get another two weeks' graze in June. By July, allow three weeks for regrowth; by the end of August, four weeks for regrowth. Keep in mind, however, that this formula also depends upon levels of rainfall and the condition of the soil.

Soil Makeup

The kind of soil you have depends on the region and other factors and dictates what types of grasses and forage will grow well. Get a soil sample test kit from your local Soil Conservation Service (SCS), extension service, or farm supplier. You must perform soil tests to assess your pasture so that you can determine the proper steps to take for improvement. The SCS or extension service will provide you with information on how to collect the samples and also on the type of forage that will grow in your area. Soil tests will tell you how much phosphorus, potassium, magnesium, aluminum, calcium, and

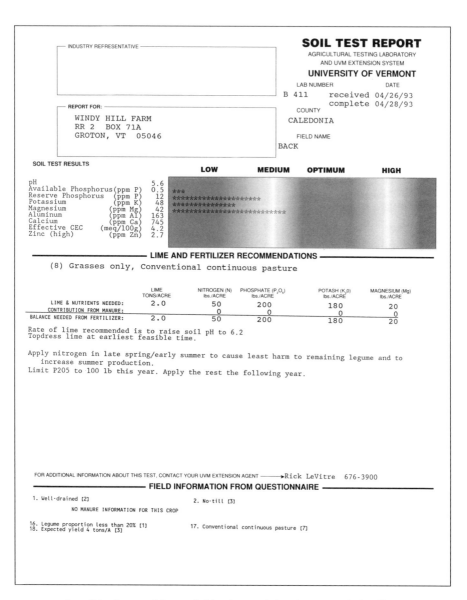

INDUSTRY REPRESENTATIVE

SOIL TEST REPORT

AGRICULTURAL TESTING LABORATORY
AND UVM EXTENSION SYSTEM

UNIVERSITY OF VERMONT

LAB NUMBER   DATE

B 411   received 04/26/93
          complete 04/28/93

COUNTY

CALEDONIA

FIELD NAME

BACK

REPORT FOR:

WINDY HILL FARM
RR 2   BOX 71A
GROTON, VT   05046

SOIL TEST RESULTS

			LOW	MEDIUM	OPTIMUM	HIGH
pH		5.6				
Available Phosphorus	(ppm P)	0.5	***			
Reserve Phosphorus	(ppm P)	12	********************			
Potassium	(ppm K)	48	****************			
Magnesium	(ppm Mg)	42	**************************			
Aluminum	(ppm Al)	163				
Calcium	(ppm Ca)	745				
Effective CEC	(meq/100g)	4.2				
Zinc (high)	(ppm Zn)	2.7				

LIME AND FERTILIZER RECOMMENDATIONS

(8) Grasses only, Conventional continuous pasture

	LIME TONS/ACRE	NITROGEN (N) lbs./ACRE	PHOSPHATE (P₂O₅) lbs./ACRE	POTASH (K₂O) lbs./ACRE	MAGNESIUM (Mg) lbs./ACRE
	LIME TONS/ACRE	NITROGEN (N) lbs./ACRE	PHOSPHATE (P_2O_5) lbs./ACRE	POTASH (K_2O) lbs./ACRE	MAGNESIUM (Mg) lbs./ACRE
LIME & NUTRIENTS NEEDED:	2.0	50	200	180	20
CONTRIBUTION FROM MANURE:		0	0	0	0
BALANCE NEEDED FROM FERTILIZER:	2.0	50	200	180	20

Rate of lime recommended is to raise soil pH to 6.2
Topdress lime at earliest feasible time.

Apply nitrogen in late spring/early summer to cause least harm to remaining legume and to
increase summer production.
Limit P205 to 100 lb this year. Apply the rest the following year.

FOR ADDITIONAL INFORMATION ABOUT THIS TEST, CONTACT YOUR UVM EXTENSION AGENT ───────►Rick LeVitre   676-3900

FIELD INFORMATION FROM QUESTIONNAIRE

1. Well-drained [2]
     NO MANURE INFORMATION FOR THIS CROP

2. No-till [3]

16. Legume proportion less than 20% [1]
18. Expected yield 4 tons/A [3]

17. Conventional continuous pasture [7]

*A soil test report is useful in determining how much fertilizer
will be needed to improve the nutrient level in the soil.*

zinc is in your soil. Laboratories generally charge between $10 and $25 to analyze a soil sample. By knowing what types of forage your pastures presently contain, the laboratory can recommend the amount of lime and fertilizers you'll need to balance deficiencies in the soil.

You won't be able to change geography, the number of foraged acres, or soil type. Your finances may limit the rate at which you can improve the quality of your forage. Getting neglected pastures into condition for proper grazing will be an investment. Depending on how poor your soil is, even liming and fertilizing can be expensive. Each acre may need as much as 2 tons of lime and 2 tons of fertilizer. The costs can add up quickly if you plan to recondition even just 10 acres. If you cannot afford to lime and fertilize to the specifications the soil tests call for, just start gradually. Whatever you add will be an improvement.

Other Concerns

Proper drainage, condition of fencing, poisonous plants, and wooded areas versus pasture also need to be evaluated. Grazing areas that are not properly drained are too wet to produce the best forage for llamas. Bogs and swampy areas should be off-limits to llamas, not only because they contain poor forage but also because parasitic problems could develop (from mosquitoes, for example). Drain a damp or wet area by installing stone drainage ditches or a culvert.

The condition of your fencing can also affect grazing. Poor fencing may not contain llamas in their present grazing areas or keep them out of growth sections. Llamas love to forage in wooded areas, but they also like to strip the bark from trees. If you have any favorite trees, keep llamas out of those areas or build protective fences around them.

Black cherry, black locust, black mustard, nightshade, boxwood, and mountain laurel are some of the common plants that are toxic to llamas. You need to be able to recognize them. *A Reference Guide to Poisonous Plants* is available from the Greater Appalachian Llama Association, c/o Linda Hoyt, RR 33, Box 2900, Skowhegan, ME 04976.

Poisonous Plants

Below are some of the more common poisonous plants. Some have been known to be poisonous to livestock, but no specifics are yet available for llamas.

Black Cherry: Highly toxic and the deadliest to llamas. Found throughout the United States. The bark, twigs, seeds, and wilted, fresh, and dry leaves are poisonous. Can cause respiratory failure, spasms, muscle twitching, and eventually coma and death.

Black Locust: Found in Canada, eastern and central United States. The bark, seeds, fruit, and trimmings are toxic. Can cause depression, weakness and paralysis, appetite loss, bloody diarrhea, and vomiting.

Black Mustard: Found throughout the United States and Canada. Chemical compounds in the seeds become toxic when mixed with water. Creates a volatile oil that burns or causes blisters. May cause abortion.

Nightshade (black): Widespread throughout the United States. Unripened berries and leaves are toxic and can cause appetite loss, unquenchable thirst, diarrhea, inability to stand and overall body weakness, and eventual coma.

Nightshade (deadly): Found in the eastern United States. The entire plant is toxic; can cause increased heartbeat, rapid pulse, and drying of mucus membranes.

Black Mustard

Boxwood: Found in southern New England and along the East Coast. Leaves are the most toxic, but ingestion of any part of the plant may cause lack of coordination, muscle seizures or tremors, abdominal pain, and vomiting or diarrhea.

Bracken Fern: Widespread throughout the United States, but limited to sandy areas; not normally found in plains states. The entire plant is toxic. An enzyme in this fern affects the metabolism, causing certain death when an animal eats its comparable weight. Death occurs within 3 months.

Buttercup

Buttercup: Found in the eastern and central United States, but not in prairie areas. The entire plant is toxic. It affects the central nervous system and can result in paralysis (in advanced cases), colic, blindness, and diarrhea.

Rhododendron: Found from Maine to Georgia, west to Ohio and south to Alabama. The leaves are toxic and can cause depression, salivation and foaming at the mouth, vomiting, and eventual coma.

Groundsel (also called Ragwort): Found throughout the United States. The entire plant is toxic. When ingested, it can cause appetite loss, unhealthy fleece and skin, cirrhosis of the liver, and coma.

Milkweed

Poison Hemlock

Henbane: Found in Rocky Mountain areas; also in the northern United States and southern Canada. Highly dangerous when fresh or dried leaves of immature plants are eaten. After fruits ripen, it is no longer toxic. Can cause convulsions, blurred vision, dizziness, and eventually death.

Horse Nettle: Found from New Mexico to South Dakota; Florida north to New York. The entire plant is toxic; causes digestive problems. Because of its thorny nature, livestock tend to steer clear, but there have been reports of problems in South Dakota.

Milkweed: Found from southern Canada to Florida, most common in the eastern United States, but also in Alabama, Kansas, Oklahoma, and South Dakota. Leaves and stems are toxic and can cause weakness, depression, convulsions or diarrhea. However, it is highly distasteful to livestock.

Mountain Laurel: Found in the eastern United States. Leaves are toxic when eaten in quantity, causing diarrhea, runny nose and eyes, appetite loss, excessive rolling, drooling, inability to stand, and coma.

Oak: Found in most regions throughout the United States. Immature acorns and wilted leaves are toxic. Can cause kidney failure or damage.

Ohio Buckeye and Yellow Buckeye (member of the horse chestnut family): Found in eastern Tennessee, central Alabama, central Oklahoma, Nebraska to Iowa, West Virginia, and western Pennsylvania. Nuts, leaves, and young sprouts are toxic. Can cause weakness, trembling or shaking, vomiting, and coma.

Poison Hemlock: Found throughout the United States with the exception of desert areas. The entire plant is toxic and will cause rapid and shallow breathing, twitching or shaking muscles, and increased weakness.

Rattlebox: Found from the eastern United States to eastern Great Plains. The entire plant is toxic; can cause gradual weight loss and liver problems. Can be fatal to all livestock.

Sheep Laurel: Found from Virginia north to Newfoundland, northwest from Georgia to Michigan. Flowers and leaves are toxic, causing drooling and salivation, dizziness, runny nose and eyes, abdominal cramps, vomiting, and diarrhea. Highly poisonous to all livestock when eaten in quantity.

Skunk Cabbage: Found throughout the United States. All parts of the plant are toxic. May cause difficulty breathing, slowed heart rate, and excessive salivation.

Staggerbrush: Found from Florida north to New York, southern Rhode Island, eastern Texas to Arkansas. Leaves and flower nectar are toxic, causing salivation, burning in the mouth, vomiting, diarrhea, convulsions, and coma. Highly poisonous to grazing livestock.

St. John's Wort

St. John's Wort: Found throughout the entire United States. Stems and leaves will cause sensitivity to light when ingested. Can also cause skin burns resulting in skin inflammation and itching; tissue will then rot, forming heavy scabs. Symptoms can be confused with parasitic infections and/or mineral deficiencies.

Water Hemlock

Water Hemlock: Found throughout the United States. The entire plant is toxic; can cause increased heart rate with rapid breathing, coma, and death.

Wild Lupine: Found on the eastern coast; also in West Virginia, Ohio, Indiana, and Illinois. The entire plant is toxic, including ripe seeds. Can cause respiratory problems and death.

Four

Facilities and Equipment

Llamas are adaptable, so there are few strict guidelines for facilities. You'll need housing and fencing, as well as work, feed, tack, and storage areas. Facilities should be designed for proper herd management and weather extremes.

Housing

Even though the llama has an impressive ability to cope with most any type of climate, some guidelines do apply.

Basic windbreaks are necessary. You can provide a simple three-sided shelter — even a free-standing shed — or a simple addition to an existing structure. In warmer climates, provide shelter from the sun. An open-air, sideless shelter may be sufficient.

Llamas should always have a dry place to lie down. In colder climates, provide bedding to keep them off the cold, damp ground. This is particularly important for older llamas with arthritis or rheumatism.

Warmer Climates and Hot Weather

Llamas are susceptible to heat stress in hot and humid climates. Even New England has days of intolerable heat. Shade and/or shelter should be available at all times. In hot climates, a more open shelter or a "loafing shed" (a place where the llamas can "escape" the weather or simply "loaf around" together), strategically placed to allow airflow from natural breezes, works well. If you can't provide this, install fans in a more permanent facility to cool those animals that are more prone to heat stress; namely, dark-colored llamas, overweight llamas, and nervous studs.

A friend and fellow breeder in New Hampshire came up with a most ingenious arrangement. Large fans placed strategically in the barn are connected to a motion detector. When the llamas enter the barn for shelter on a hot day, the motion detector turns on the fans automatically. When the animals leave the barn, they trip the motion detector again and the fans shut off.

Sprinklers on a timer will offer a cool break during the hottest part of the day. This may sound expensive, but if your facilities are small, as most start-up facilities are, the investment of two or three 50- or 100-foot hoses and a "fan-type" sprinkler is minimal. Timers run from $10 to $20. My farm is located in colder Vermont, but I use this system on hot days. I have seen my llamas head for the sprinkler and lie down next to it 5 minutes before it is scheduled to turn on. The internal clock these animals possess is amazing. Your llamas will quickly learn when the timer is set to turn on the spray.

Fresh water should be readily available. Llamas drink less water during the grazing months because they derive most of their moisture from the forage. During months of supplemental feeding of hay, however, llamas need additional water because of the dry forage.

Colder Climates

Llamas will lie on the ground with their backs to the wind to wait out snowstorms, but if shelter is offered, they will seek it. Colder northern climates call for a more substantial structure but this still can be as basic as a simple three-sided shelter. The opening should be to the south or southeast; most weather conditions in colder climates arrive from the northwest. With protection from the northwest wind, the south-facing opening will allow a sunny exposure on clear days.

Size of Shelter

Geldings (castrated males) and young llamas do not need a strategically placed shelter to keep them separate from other llamas. A simple three-sided shed with a well-bedded floor of straw about four inches deep will suffice. If you plan to develop a breeding operation, you should make more specific plans regarding the placement of shelters and fencing, keeping expansion in mind for the weaning of babies, accommodation of studs and open females, and other management issues. This will create more work in the beginning, but you'll be glad of it in the end.

Specific plans for weanlings, for instance, should include an area with shelter several pastures away from mom to lessen the possibility of attempted escapes. You want weanlings out of sight of mom but in the company of other llamas. In the case of studs, depending on their aggressiveness, they should be housed independently in any operation.

Some larger farms house their studs together, but still completely removed from the females. The females should not even be in sight, and should be upwind if possible.

If pasture breeding is utilized, open females that are not being bred need to be kept separate from the breeding stud. (Open females are females that are not presently bred. In pasture breeding, the sire runs continually with the open female until she becomes receptive to the male.) Open females should not be housed near breeding studs unless the stud is fairly docile and highly manageable.

The size and location of housing will depend on the ultimate size of your herd. It is always easier and less expensive to build with expansion in mind than to add on later. Also keep in mind that llamas are very social and as many as eight to ten llamas will fit into a stall roughly 8' x 8'. They will happily lie with their bodies touching. (For the most part, llamas are adaptable to any type of housing as long as they are free to come and go.)

You can erect a three-sided shelter for as little as $300. A loafing shed may be a three-sided shelter either attached to a barn or standing alone. It is simply a place where the llamas will congregate on hot or cold, windy days or just to socialize at any time.

Gale Birutta

A horse barn is ideal for llamas. As many as five llamas may share a single horse stall.

Fencing

Fencing can enhance your property's beauty and value. You can choose high-tech and expensive options such as PVC (a type of plastic fencing, made from the same material as plastic piping in plumbing) or low-cost fencing such as cedar posts with insulators and smooth electric wire. Whatever your taste and budget, there are two components to consider: confinement and aesthetics.

Llamas are respectful of fencing. But a llama can easily jump a 5-foot fence, and one that is lonely or looking for food will often make the attempt. Highly social animals, llamas do not like to be alone and will seek a way out to join other llamas or livestock. You can confine llamas with a 4-foot fence of any type; requirements are minimal. Among your options are split rail, high tensile, portable sheep fencing, and board fencing.

The height of your fence depends on the sex of the animals and their access to forage. In most cases, breeding studs will require a more substantial fence (at least 4 feet high) when in close proximity to females. (The grass on the other side of the fence won't tempt a llama if there is adequate feed and water within his own area.)

Management concerns such as breeding studs, open (nonpregnant) females, and weanlings must be considered. Generally, it is better to have a higher fence between studs and open females, as well as fencing that is crawl-proof for weanlings.

Llamas are aggressive against predators, but dogs or wolves can take babies if their dams are unable to defend them. However, a neighborhood dog that

Types of Fencing

Type	Pros	Cons	Suppliers
Board	Attractiveness; availability; strength; durability	High maintenance; costly; time consuming to install	Fencing contractors; ag/farm stores
PVC	Attractiveness; durability; no maintenance	Most costly; time consuming to install	Fencing contractors
Split rail	Attractiveness; durability; strength; availability	Costly; escapability between rails; needs added strands of electric wire	Ag/farm stores; building suppliers; fencing contractors
High-tensile electric	Affordable; easily and quickly installed; availability	Entanglement without escape	Kiwi Fence; fencing contractors; ag/farm stores
New Zealand Spider electric	Highly affordable; easily and quickly installed; highly adaptable to fewer changes in terrain; fence posts	Entanglement, but allows escape without serious injury; available only from dealer (Kiwi)	Kiwi Fence
Woven wire	Affordable; easy to install; easy maintenance	Entanglement	Ag/farm stores; hardware stores; building supply
Portable sheep	Easily moved; reusable; longevity	Can be costly; available only in 32–37½" heights (too short)	Ag/farm stores

accidentally finds its way into a dam's pasture when she is with her cria is in for a surprise. Female llamas are very protective of their young and will put up quite a fight. Packs of dogs, however, can present a challenge and a very real danger. The dam will become confused and not know where or how to defend her baby. Llamas are no match against packs of wolves, either, because of the wolf's cunning nature. Wolves work in a coordinated team effort. Some distract the dam while the others take the offspring. A female llama will have trouble protecting herself as well; wolves are one of the few species that llamas can fall victim to. Therefore, any type of wood fencing with wide openings between the rails or boards would need special consideration, particularly in areas where domestic or feral canines run. Canines can slip through boards and rail fencing easily and make their way into the pastures. An electric wire would have to be run between the bottom rail and the ground to keep them out. The predator will usually back off as soon as the shock is felt.

Barbed Wire

Barbed wire is strictly off-limits for llamas. Remove it if it already exists on your property. As people become educated in modern-day types of fencing, barbed wire is gradually disappearing from many farms. Animals can seriously injure themselves when frightened or seeking food. The barbs on the wire can cause significant wounds if the livestock becomes entangled.

PVC Fence

The PVC fence is a high-cost alternative to board fences. It offers the look and prestige of the board fence but is made of a hollow-core PVC material. It will beautify your grounds and won't require any paint, but it will cut into your budget significantly. Four feet is the standard fence height for livestock, and that's true for housing llamas, too. Four-foot fencing will adequately house breeding studs (they present the biggest challenge to any type of fence). You can get 4-foot PVC, but if you don't plan on housing breeding studs, a sheep containment height of 3 feet will be sufficient.

PVC fencing eliminates maintenance.

Board Fencing

The most attractive fencing is the board fence, more commonly known as the Thoroughbred fence. These are the white-washed fences you see lining the borders of Kentucky Thoroughbred farms. These beautiful but costly fences offer durability and strength. They're a high-maintenance option, requiring painting or staining.

You can accomplish the same wood effect by using pretreated lumber. Pretreated lumber is usually guaranteed for at least 30 years and should last up to 50. Some chemicals used in the treating process are toxic to livestock if swallowed, but the smell is enough to keep most livestock away.

Llamas do not chew or crib on fencing and lumber as horses may do. Llamas have a hard upper palate with no teeth in the front; thus, chewing on wood, fences, and gates is simply not possible.

Split-Rail Fence

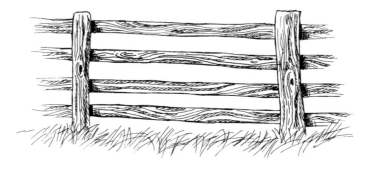

Split-rail fences should have no more than 10 inches between horizontal rails so the llamas cannot squeeze their heads through.

The split-rail fence is another attractive wood fence. The important consideration with split-rail fencing is that the rails must be set "horse-style," which means that the horizontal rails should not exceed 10 inches apart in width. Any more than that might allow llamas to get their necks through. (A smart llama will squeeze his way through.) In the case of babies or weanlings, you may need to add a single strand of electric wire for additional security between the bottom two rails. Or you can substitute woven wire to add protection. With a split-rail fence, this is almost a requirement for young animals because of their susceptibility to predators.

High-Tensile Electric Fence

Llamas can be kept in an electric fence with just a single strand. Provided they are in a rotational grazing program with enough forage, most females and geldings will respect a one-strand electric wire. Be aware, however, that some may not; you may need to use more than one strand.

Some larger and more commercial farms utilize high-tensile fencing. This is a system of pressure-treated posts set 4 feet into the ground. The end posts are firmly and securely braced to carry the high-tension, 12-gauge wire. High-tensile fencing is popular in the llama industry because it's affordable and quick and easy to install. Wires are closer together than in other fence systems — from 3 to 6 inches apart — with a ratchet-type tensioner used to keep the wire taut.

This fence is both a mental barrier with an electric charge as well as a physical barrier. Some problems have been reported with this fence: Crias or weanling llamas, in an attempt to escape, have proceeded halfway through and then gotten their hips and hind ends hung up on the taut wire. This could lead to disaster — if circulation is cut off for an extended period of time, it may be necessary to amputate a limb.

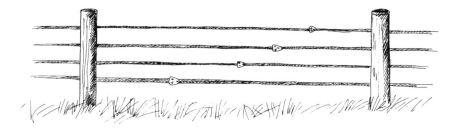

A high-tensile electric fence provides both a mental and a physical barrier.

New Zealand Spider Electric Fence

Although relatively unknown in the United States, this system was developed to accommodate extensive rotational grazing in the rocky and mountainous areas of New Zealand, where conventional fence posts can't be set. This fence requires only well-set corner posts and interim fiberglass posts. (When setting the posts in rock, a portable gasoline drill with a carbide tip works well. Set the rod 4 to 6 inches into the rock and secure it with a couple of handfuls of concrete mix.) This versatile system can be installed quickly and easily to enclose heavily wooded and wet areas, or for normal pastures. It's available from Kiwi Fence dealers (see Appendix B).

This fence has a powerfully potent charger with up to a 17.0 joules rating to deter predators and confine animals. This is a flexible fence, and livestock are seldom injured even if they become entangled in the wire. The manufacturer reports no serious injuries to livestock other than an occasional small cut to cows and horses. Under heavy pressure, the wire is designed to stretch and even break to avoid injury, but livestock quickly learn to avoid the charge. Llamas do not thrash around and are generally smart enough to stay

Vermont's Weather and Llamas

Vermont has some interesting weather. When the cold air from Canada clashes with the warmer southern air, anything can happen. It was a warm May evening, and the cool Canadian air was pushing down in along the mountains. I was alone in the house. There were some faint rumblings of thunder in the distance — about 10 miles away. No wind, no rain. I looked out the dining room sliding door for the last visual check of the evening and then proceeded to the living room for some quiet reading. As I passed by the doorway to the cellar stairs, a deafening blast and a blinding flash pushed me back, knocking me to the floor. As I picked myself up, I thought that something had exploded in the house, perhaps the propane water heater. In a few seconds, I realized that the house had been hit by lightning. My first thoughts turned to the llamas, as our farm is all open pasture situated on the side of a hill with a large pond.

The power was out and I made my way into the cellar for a flashlight. Something was burning and I was getting nervous. I tried to call the fire department but the phone was dead. There was no rain, no thunder, no lightning. I ran between the paddocks to the neighbor's and did a llama head count on the way. There they were, all as peaceful as ever, oblivious to the confusion around them and quietly chewing their cuds.

My neighbor was also without power, but he came back with me to check things out. I hadn't thought to look at the electric fence energizer, but there it was, blown right off the wall in the cellar, still smoking. Dozens of pieces were strewn about the basement. Someone from the power company arrived about two hours later and realized the transformer on the pole would need replacement. The pole was about 5 feet from one of our paddocks. The representative from the power company said he had never seen a transformer in such a condition. Evidently, a distant lightning hit caused a ground surge that followed our electric fence directly to the line into the house and found its way out through the charger. Luckily, the manufacturer guarantees its chargers against lightning strikes, so we were able to get it replaced. And even more luckily, the llamas were unharmed and unfazed.

calm and quiet until help arrives. This fence is highly effective and above average in safety for the containment of llamas.

You won't need heavily secured end post bracing or ratchet-style tighteners. G-springs take up any slack that occurs, and they also double as gates. Additional posts may be set to accommodate gates. G-springs are made of tensioned steel. Under heavy tension, these springs will actually stretch enough and flatten out to allow additional slack to the fence, thereby averting serious injury. (On million-acre New Zealand sheep farms, skids are installed on the front of vehicles to allow the farmer to drive right over the fence without having to search for gates. The G-springs regain their shape immediately and automatically spring the fence back up to its normal position.)

The New Zealand Spider electric fence is an adaptable, inexpensive, and eye-appealing fence. It has been relatively slow to catch on in the United States because it is so different. People still tend to believe that fencing needs to be a physical barrier. This fencing is strictly a mental barrier.

It is still considered portable fencing by most zoning ordinances and therefore is not usually taxed as a property improvement. With the exception of portable sheep netting, all other types of fencing are deemed a property improvement and therefore are taxed. This will vary from state to state and town to town — it is best to check with your local tax assessor.

Fiberglass posts follow the contour of the terrain and can be set in any type of ground including shale, sand, muck, and even running water.

Woven Wire

This fencing is available in 3- and 4-foot heights, with rectangular openings that range from 7 inches × 8 inches at the top to 3 inches × 8 inches at the base. Although many llama breeders utilize this type of fence because it is inexpensive, quick to install, and easy to maintain, there have been some disasters. Crias have become entangled up to their hips, and some have been seriously injured, even paralyzed. (Others have been extremely lucky and suffered no permanent injury.) While it will continue to be a commonly used fence on farms because of its low cost, it does pose a problem to young animals and is not recommended for llamas. When housing adult llamas only, it can be used effectively; if you choose this fencing, the rectangular openings should not exceed 3 inches.

Support this fence with corner posts, and with interim posts set 8 feet apart. The life of this fencing can be up to 15 years, but of course this depends upon the type of posts, the depth to which posts are buried, and wear and tear.

Gates

Gates can be strategically placed as shelters. Llamas like to gather at gates. (They seem to think that the llama closest to the gate will be the first one fed.) Gates are usually better when they are designed to open *into* the containment area rather than out. Overzealous llamas may push their way out during feeding.

Breeding males tend to mark a gate as their "territory" by depositing their dung pile there. There is really no good solution: A breeding male quickly learns that the females either enter or depart via "his" gate.

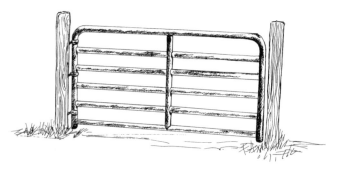

You can purchase aluminum gates at most farm, ranch, and feed supply stores. They start at under $40 for a 4-foot gate and go up to around $100 or more for an 8-foot gate.

Gale Birutta

Wood gates are attractive and easy to make.

Be sure to latch gates securely; llamas can be escape artists. Kiwi Gate Latches are widely used in the llama industry as they self-lock by way of a ring that drops into place and cannot be maneuvered out of place by livestock. (See Appendix B for suppliers.)

Gates come in many varieties. You can purchase them at the local feed store or make them inexpensively at home. You'll find gates in almost any size to accommodate your needs.

Equipment and Tools

A new llama owner will need certain equipment and tools. You'll want restraining chutes to contain difficult or unmanageable llamas, scales, toenail trimmers, grooming supplies, shears, and brushes. You can purchase tools and supplies through mail-order companies, or try your local hardware store or general farm and garden supply store. Compared with the high-priced supplies needed for other livestock ventures, you'll be surprised at the minimal cost of these items. (See Appendix B for a list of suppliers.)

Chutes

Restraining chutes make it easier to handle difficult llamas. If you have done your homework and purchased manageable animals, then you won't need a chute immediately.

Proper restraint for some veterinary work is a must. Some llamas may be more unruly than others. Toenail trimming, the removal of fighting teeth, and even shearing may require a restraining chute. Larger breeding operations will ultimately require a chute, as it will make handling of each llama safer in general for the breeder. Most larger llama supply houses carry ready-made chutes that you can assemble quickly. Or you can build your own wooden chute. Plans are available from suppliers listed in Appendix B. These chutes are adjustable to all sizes of llamas and the cost of materials can run anywhere from $50 to $150. Pre-made or manufactured chutes can cost between $250 and $1,000.

Llama restraining chute: You can build your own or buy from a supplier.

Chute manufactured by The Llama Connection, distributed by Quality Llama Products, Inc.

Scales

A large farm or breeding operation needs a walk-on platform scale. A regular weighing program for all your animals is important to guarantee proper nutrition balancing (see Chapter 3). You'll be able to detect problems such as overconditioned and underconditioned (overweight and underweight) animals. Scales will also help you to identify crias that fail to gain sufficient weight. Platform scales range in price from $350 to $1,000. It may be possible to locate a used one more inexpensively from a livestock producer who is retiring from the business.

Special cria scales are available. These are simply hanging scales, from which a sling can be made to fit under the cria's belly. Hanging cria scales cost between $60 and $85, and will accommodate crias up to 60 pounds. An ordinary but accurate bathroom scale is also useful. Crias are light enough to pick up and carry onto the scale.

Cria scales measure in 1-ounce increments, so you'll get a very accurate account of the baby's weight gain.

Kevin Kennefick

Toe shears

Kevin Kennefick

Hoof nail nippers

Toenail Clippers

Great news for sheep breeders reading this book: Llamas do not get foot rot, nor can they catch it from sheep if housed with them. In sheep, foot rot is caused by a bacterium that thrives in warm, damp weather.

Hoof nail nippers and toe shears are the toenail clippers most widely used by llama breeders. No one toenail clipper can be recommended, however, as llama owners have used everything from pruners to blacksmith nippers. Three or four types of nail clippers are available from farm stores or suppliers listed in Appendix B of this book. You may want to try several types to find the one that works best for you. I have found the shear-type nippers to be the best at a cost of about $15. They are more versatile than other types and can also be used on sheep and goats. Other uses include cutting baling twine or light wire.

Blacksmith nippers are cumbersome and difficult to use. Toenail pieces also become struck in the blades and need to be removed before continuing to trim. This is hard when trying to hold a llama's leg with one hand and the nippers in the other.

Utility knives should never be used because of the possibility of slipping and seriously cutting your llama or yourself.

Groomers and Blowers

Tools for basic grooming include brushes, slickers, and rakes. Rakes are excellent for everyday use as they go through the wool easily with less stress on the llamas. A slicker with soft wire is also good for everyday use. Grooming llamas depends on your time schedule and llama use. Pack llamas should have a quick grooming to remove any debris before the tack is applied. 4-H llamas can be groomed daily for manageability of the animal. Alpacas should never be groomed on a daily basis as it damages the fleece.

Kevin Kennefick

The Circuiteer II is a high-volume blower used for grooming llamas.

If you are planning any serious showing, I recommend that you buy a high-volume blower. They cost several hundred dollars, but they're worth the expense. Even if you are not showing, this piece of equipment can be used to desensitize the most uncooperative llama. After shearing, the high-speed blower will relax your llama and promote better circulation. Refer to Chapter 13 for more information on fleece.

Shears

The best hand shears are English-made, and are available from mail-order or feed and livestock stores. I prefer the 3½-inch shears originally developed for sheep. These shears are also available with 6-inch blades, but their larger opening makes them harder to hold; however, they will get the job done quicker. The cost of 3½" shears is about $17; of 6" shears, about $25.

Electric shears can be quite expensive, ranging from $200 to $400. When these are first used,

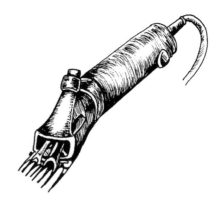

Electric shears should be used with offset blades that do not shear all the way to the skin.

Kevin Kennefick

Standard 3½-inch shears are popular among owners and breeders who regularly shear their animals.

some llamas do not like the sound and may be impossible to shear. A llama's fleece is also an insulator against sun and heat, and close electric shearing can cause sunburn on light-colored llamas. Llama suppliers offering electric shears have "offset" blades available, which do not shear to the skin. Be sure to ask for 1½-inch length blades.

Tack

Basic tack requirements for llamas include halters, lead ropes, and picket and stake out lines. If you are planning to pack your llama, either recreationally or commercially, see Chapter 12 for more information on the tack needed for this.

Halters

Because llamas are cud chewers, they need a specially designed halter. A properly fitting llama halter should follow these guidelines:

- ◆ The noseband of the halter should be *at least* 1½ inches above the end of the nosebone. If the noseband of the halter is too low on the llama's nose, it will cut off his airway.
- ◆ You should be able to fit a medium-sized hand between the underside of the jaw and the halter ring at the bottom.
- ◆ The strap that is buckled to attach the halter should not be buckled so high up or tight that the noseband rises up on the llama's face. The strap should also not be buckled so tight that it interferes with the llama's eyes. The edges of the halter should be at least one inch from the llama's eyes.

Many people think pony halters will work for a llama, but these halters do not allow enough room under the jaw for cud chewing, and the way they are cut they will interfere with a llama's eyes. Avoid any halters that are not

Haltering a Llama

1. Approach the llama with the halter around your forearm with the lead rope attached.

2. Wrap the lead rope around the llama's neck to hold him.

3. Slip the nosepiece over the llama's nose.

4. Buckle the crown of the halter behind the ears.

5. You are ready to lead your haltered llama.

specifically made for llamas. Any llama supplier has several types available. Llama halters are available in nylon or leather, ranging in price from $6 to $12 for nylon, and up to $18 for leather.

Choosing the Right Halter

Standard halter	These are best for general use.
Fixed ring halter	These offer maximum control and are good for early training or hard-to-handle animals.
X-style halter	These provide greatest flexibility for grazing; they are not good for training.
Adjustable halter	These come in handy if you don't know what size you will need, or if you want one halter to fit several llamas.
Sheepskin halter	These are made for llamas that have tender noses or that may have developed sores in the past.

Do not leave any halter on your llamas fulltime. The halter can rub and cause sores, or can get caught in fences or brush.

Lead Ropes

Lead ropes are supplied by all the llama supply houses but are also readily available at most feed and grain stores. Lead ropes may be of poly pro (a petroleum-based material) or pure cotton. Cotton leads are preferred because of their softer feel. Available in a wide array of colors, they range in price from $3 to $8 each.

Picket and Longe Lines

A picket line is used to graze your llama when not in an enclosed field or pen. This is very useful on the trail during packing. It involves the same concept as tying out a dog on a long lead. Screw-in stakes are available at hardware and feed stores for about $4 each. A longe line can be attached to the stake to allow the llama to graze and browse. Horse longe lines are available from horse suppliers at a cost of about $20, depending on quality. These will allow the llama to graze up to 25 feet. Two 8-foot leads may also be tied or permanently attached together for up to a 16-foot line. We have made our own with climbing rope and attached a clasp to both ends. A swivel at both ends helps to keep the line from twisting and prevents legs and feet from becoming entangled.

Dealing with Emergencies

When you keep safety considerations in mind, you will avert most emergencies and disasters. But things do happen. Tragedy and financial loss can be kept to a minimum by being prepared.

If you are a new owner, it is important to think about housing or a holding area in case of an emergency. Consider close-by alternative housing and containment areas.

When you are the guardian of animals, you must have plans that ensure their safety in any circumstance. Trying to stuff several llamas in a horse trailer at the last minute and having to move and unload them to go back for the others simply does not work. You could lose animals without planning and foresight.

Having a good, working relationship with your llamas will benefit you greatly during emergencies. The key is to stay calm. A llama will instantly pick up on your anticipation, fear, or excitement and will become agitated. Keep in mind that llamas are naturally head shy, and trying to put anything over their head or eyes (like you might do with a horse) will surely result in panic and instant unmanageability.

Evacuation

Have a plan in case you must evacuate your animals. The most common reason for evacuation is fire, which can be caused by, among other things, internal combustion in the hayloft. There may be an earthquake, a flood, forest fires — your regional location will dictate what natural disasters you must prepare for. In all cases, you must have a plan made well in advance. When warnings of disaster are issued, don't wait to see how close the disaster will come; *move immediately!*

Building Fires

With almost all building fires, it is best to maintain a "holding area" nearby but generally away from other buildings and fencing. With a rapidly moving forest or brush fire, you'll want to evacuate the animals completely from the premises.

Llamas dislike being locked in; they want to come and go as they please. This may work in their favor in an emergency. If their paddock is large enough, you will be able to remove animals from a structure fire immediately by moving them to the rear of the paddock until they can be evacuated completely and safely from the site. Make sure they are secure in their temporary location. Llamas can become highly stressed and confused in an emergency situation. If left on their own with no guidance, or if they are not handled calmly, they'll seek a natural and familiar place and may return to the burning building.

Forest and Brush Fires

If your farm is in a region with dense vegetation, always beware the threat of wildfire. Fast-moving brush and forest fires caused by heavy winds are nearly impossible to contain — you must be prepared for complete evacuation of the premises.

Some preventive measures may avert or lessen a dangerous situation. If you are in a fire-prone zone, enlist excavators and the help of the forest service to install fire breaks around the property. Prior preparation goes a long way. The recent California wildfires wiped out many farms and livestock. Don't wait until a fire is headed your way. Make sure brush and combustibles are kept at least 50 to 100 feet from any structure where animals congregate or are housed. Unfortunately, no matter how well prepared and how well rehearsed we are, emergency situations are emotional. Try to stay calm and clear-headed — don't allow panic to take over under stress. A well-planned evacuation that has been reviewed with your family and any staff will greatly increase your chances of removing all animals safely.

If your farm has numerous animals, you need to think seriously about the amount of transportion you have access to in case evacuation is necessary. You may have to prioritize your herd, removing the most valuable animals first. Don't wait until it is necessary to make split-second decisions about which animals to take. Plan ahead!

If you are a diversified livestock grower, you will have to make your own priority list of which animals are of the highest value to you. In all cases, any species utilized as breeding stock are valuable. Depending on your business, the livestock that generates income for you should be removed first.

If you are in a high-risk area such as a wildfire, coastal, or flood zone, or in a hurricane or tornado area, a drill involving your family, staff, and animals is MANDATORY at least once per month. Other less high-risk areas can hold drills perhaps two times per year.

Contact your local civil defense, fire, or rescue department. They will be happy to work with you in creating a "mock" drill or disaster. This keeps their skills sharpened as well as your own. This is an excellent opportunity to work with rescue units and is highly recommended.

Hurricanes and Nor'easters

Farms located near the eastern coast and on large inland waterways must be prepared for flooding and/or high winds from major storms. It is imperative that buildings be constructed to withstand these hurricane-force winds. Extra reinforcement for flood-prone areas includes sinking ground posts up to 5 feet, or using block or poured-concrete foundations. High-wind areas require the same stable foundation structure and roofs constructed to local specifications for your weather conditions. In the case of rising floodwaters, construct a holding area on high, level ground for evacuation purposes.

Other Disasters

Various other disasters, both natural and man-made, may test emergency measures. Tornadoes, earthquakes, and extreme heat and cold are other conditions that you must prepare for.

Emergency Checklist

1. **Plan ahead.** Discuss the details of your evacuation plan with members of your family, farm managers, and the local fire department.
2. **Have alternative containment facilities *on the property*.** "On the property" is stressed here because you will save precious time if you can simply move some llamas and then return for the others. Again, these containment facilities can be used only in case of a building collapse or building fire, circumstances when the animals will not be in danger elsewhere on your property.

   Forest fires, earthquakes, tornadoes, and other storms require the animals to be completely evacuated from the property. *All animals must be out the first time, because there may be no time for a second chance!* If your facilities are not large enough for a secondary holding area, then make arrangements with your neighbors ahead of time.

   A holding area separated into several sections will prevent further confusion when placing llamas in a safe place away from the disaster. Try to separate studs from each other and from

open or young females. Think carefully about the number of animals you have and how they would need to be contained should
evacuation be necessary.

3. **Have on hand a halter and lead for every llama, including babies.**
   While crias usually follow their dams, they become disoriented
   during mass confusion and can be separated from their mothers.
   Color-keying halters for ages and head sizes enables you to find the
   proper halter quickly. (All llamas must be halter broken as early as
   possible.) Crias as young as two months may be halter broken.
   Crias younger than this are still small enough to pick up and
   remove. NEVER, NEVER rely on unhaltered or loose crias to
   follow their mothers. They panic easily.

   The halters should be stored in an accessible place near the
   stalls or pens, not packed away in a trunk. If alternative containment facilities are impossible, then you'll have to turn your animals loose. Make sure all halters have your farm's name and phone
   number on them!

4. **Have available a llama first aid kit to deal with any injuries your
   animals sustain in an evacuation.** Supplies in a first aid kit should
   include:

     - A topical antibacterial ointment for cuts and scratches
     - Gauze wrap or bandages
     - Vet wrap
     - 7% iodine solution or wound spray
     - Scissors
     - Latex gloves
     - Digital thermometer
     - Syringes
     - Rubbing alcohol
     - Wound ointment
     - Stethoscope

   This is an excellent first aid kit to have on hand. Make sure it is
   portable; a zippered Cordura nylon bag or standard luggage
   makeup case is best. First aid kits are available through suppliers
   listed in Appendix B.

5. **Be sure your animals are manageable.** In dealing with an evacuation, the manageability of your animals may determine their fate.
   When seconds count, you want to know that they are catchable and

can be quickly haltered and led to safety.

6. **Keep your llamas' immunizations up-to-date,** particularly clostridium C and D, tetanus, and rabies. This is a must in any operation and helps to ensure that your animals have extra protection in case of water or feed contamination.

7. **Carry adequate insurance coverage.** Be sure to examine your policy carefully and know what it covers. Purchase extra insurance to cover disasters that might occur in your region.

8. **Have a plan for alternative means of trailering or transportation.** If you must move a large number of animals, make arrangements for adequate transportation ahead of time. When any natural disaster or weather condition approaches, make sure these vehicles are on standby.

9. **Get a first aid kit for yourself in addition to the first aid kit for your animals.** You may make this yourself or purchase one in a pharmacy for as little as $10. In addition, get together an emergency kit consisting of the following:

   - Flashlight with extra batteries
   - Heavy gloves for possible removal of timber or other obstacles
   - 2-way radio (walkie-talkie)
   - Map of the local area (topographical)

A poorly planned and executed evacuation can have devastating consequences. Be prepared!

FIVE

HEALTH

Llamas are hardy animals. They aren't prone to general sickness and injuries. Herd health programs will depend on geographical region, number of animals, types of feed, housing, and contact with other types of livestock in pasture. Record keeping for each individual llama is important. Update information as needed. Record weight, vaccination and worming data, reproduction history if it is a breeding animal, miscellaneous injuries or illnesses, and teeth and toenail upkeep. Include pedigree information and any special feeding requirements.

This chapter is only intended to act as a guide and is not meant to encourage you to act as your own veterinarian. As you become more familiar with llamas, you will be able to take on many of these veterinary responsibilities. But a new owner should find a veterinarian who already treats llamas or is enthusiastic about learning.

Finding a Vet Who Works with Llamas

It is important to find a veterinarian who has experience working with llamas. Consult the International Llama Association (ILA), which maintains a hot line to access veterinarians. Also check with your state or regional association; they can refer you to individuals in your area. You might even ask other breeders from whom you have purchased llamas.

See Appendix A at the back of this book for addresses and phone numbers of the ILA and state and regional associations.

Conformation and Soundness

Conformation refers to the structure of an animal; soundness is its health condition. *Sound* means free from disease, illness, parasites, and injury. The ability to spot and evaluate proper conformation in an animal comes with experience. It can be difficult to evaluate a moving animal. Many llamas can appear to be conformationally correct while standing, but the true test is to observe how they move. This requires an experienced eye.

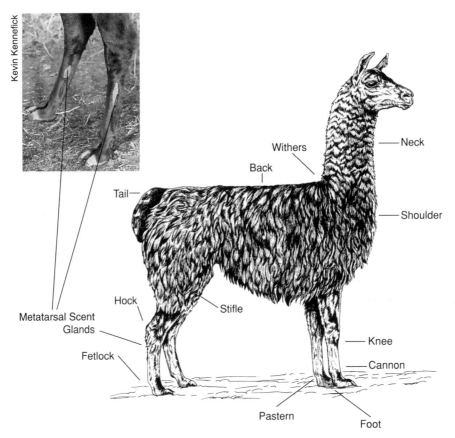

The main body parts of the llama.

Determining Proper Conformation

Originally bred as work animals, llamas must possess the proper conformation for work. Proper conformation for a llama can be compared to that of a horse. The set of the feet and legs, the slope of the shoulder, and the length of the back and its coupling relative to the length of the belly all determine proper movement. While some conformation problems can be nutrition-related such as angular limb deformity or stunted growth (see Chapter 3), most are genetic and passed from the sire and/or dam to their offspring. For this reason, it is important that breeding animals and their blood relatives be as free of defects in conformation as possible.

Correct Leg Set

Correct leg set refers to a leg under each corner of the animal. Facing the front of the llama, proper leg set can be determined by a plumb line dropped from the point of the shoulder to bisect the knee, cannon bone, ankle, and foot.

Facing the side, a line dropped from the arm should bisect the forearm, knee, cannon, and ankle on the front of the animal and pass just behind the back of the heel from the tailset in the rear. Facing the side, a line dropped from the buttock should be in line with the back of the hock, cannon, and ankle, falling just to the rear of the foot.

Facing the rear, a plumb line dropped from the point of the buttock should bisect the thigh to the hock, cannon, ankle, and foot.

Poor Leg Conformation

Poor leg conformation in llamas is termed knock-kneed, sickle-hocked, or cow-hocked. Poor leg conformation can undermine a llama's ability to pack under load. In guardian llamas, leg problems may prevent them from adequately moving at high speeds. Llamas with leg conformation problems should not be bred, whether male or female. Males must be castrated and females culled. These llamas can continue to lead useful lives as pets, companions, 4-H projects, or guardians.

Cow-hocked   *Sickle-hocked*   *Knock-kneed*

Pasterns (Ankles)

The joint of a llama's ankle should be almost vertical and not drooping toward the ground. A conformation flaw in llamas is the dropped ankle, commonly known as "down in the pasterns." This may be caused by gelding a male at too early an age, combined with overloading during packing. It is also quite common in older females that have reproduced for many years. Because the majority of a llama's weight is carried by the front legs, obese llamas may also experience this problem as they age. A packer that is experiencing dropped ankles should be retired from work. Guardian llamas with dropped ankles should be able to perform normally, provided they are in a situation that does not require extensive athletic ability.

Dropped pastern

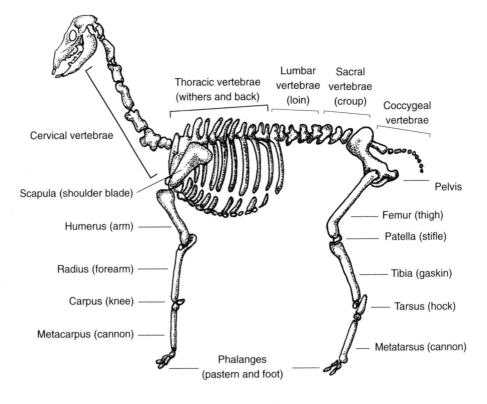

A skeletal detail of the llama.

The Alpaca's Conformation

There are distinct differences in physical characteristics between the llama and the alpaca.

The alpaca's hindquarter leg structure is different from that of the llama. The alpaca's pelvis is slightly rotated, creating a more rounded hindquarter set. This is normal in alpacas, but it is possible for alpacas to be conformationally under themselves. A llama's hindquarter leg is more in line, similar to a horse's. The normal tailset for alpacas is down from the topline. A llama's tailset is at the base of the topline.

Adult alpacas stand 30 to 39 inches at the withers and weigh 100 to 170 pounds.

Straight, rounded ears are a characteristic of alpacas. A llama's ears are longer, often banana-shaped.

Llama   *Alpaca*

Llama   *Alpaca*

General Herd Health

If you are active in the industry, you increase your llamas' exposure to parasites and diseases. If you are accommodating females for breeding services, or if you broker llamas, attend shows, competitions, fairs, and other functions, or purchase new stock, the chances of exposure increase.

By following a regular veterinary program and requiring that any outside llamas entering your farm be fully immunized, you keep problems of this sort in check. The incidence of localized disease is controlled by having animals coming to your farm from out-of-state or another province accompanied by a veterinary certificate. (This certificate is required for interstate transport.) But not all harmful conditions are covered by interstate transport certificates; ask your veterinarian to advise you.

To minimize the chances of a serious outbreak of disease, isolate any new animal arriving at your farm, be vigilant for symptoms of illness, and keep conditions sanitary.

The tidiness of your facilities affects general herd health. Manure should be raked up and removed from grazing areas at least once per week. Barn, stall, and shed facilities where llamas generally bed down should be kept clean, with fresh bedding and manure removed daily. If your water is excessively hard, buckets will develop a scum from mineral deposits that will need to be scrubbed out. I use five-gallon buckets for water. I empty them, scrub them, and fill with fresh water daily at all times of the year, whether the bucket is full of water or not. Large tubs or old bathtubs are difficult to keep clean and are breeding grounds for mosquitoes. These are fine, but be prepared to empty, scrub, and fill them on a regular basis.

Feeders are best when made of rubber, steel, or plastic. As with watering facilities, they need to be kept clean and scrubbed. Wooden feeders are difficult to clean and harbor bacteria. It is better to stay with something that can be cleaned readily.

Parasites

Llamas are susceptible to several species of parasites. How you control parasites depends on the number of llamas, pasture rotation, geographic area, confinement areas, and specific groupings of individual animals. There are two types of parasites: internal and external.

Understanding the Life Cycle

Generally, internal parasite species (worms) live in the stomach and intestines of the llama and pass out their eggs through the feces. In most species of worms, the eggs then mature to larvae, which live in the grasses and are ingested by the llama while foraging. The larvae invade the digestive tract, where they will develop into mature worms in 21 days, and the cycle continues.

External parasites are less of a problem for llamas than they are for other species. Llamas love to roll in the sand or dust. This is their instinctive way to rid their body of external parasites. Llamas will quickly make a "dust bowl" in your new pasture for this purpose. I personally have never seen a tick or lice on any llamas. While it does probably exist in regional areas, it is not

The Sick Llama

Llamas that are ill do not normally lose their appetites, unlike other live-stock species. In order to determine when your llama is ill, you must be observant and able to recognize certain signs:

- ◆ A llama who is hump-backed with his head down while standing is ill.

- ◆ A llama who lies on his side or rolls excessively may be ill.

- ◆ Llamas normally rise from a kush position immediately when approached. Llamas that do not rise could be ill or injured.

common. Injectable Ivomec (ivermectin) will alleviate most external and internal parasite problems when used as part of a regular worming program. Treatment of biting lice, however, requires a powder or external treatment (see below).

Parasite Control

Having fewer animals on a particular tract of pasture will mean less feces, and in turn fewer eggs and hence fewer larvae. Proper pasture rotation will allow sufficient time for the larvae to die before they are ingested and can continue their cycle. If weather is damp and cool, stomach larvae living in a pasture may survive from 5 to 90 days.

Overgrazing of pastures and forage forces the animals to crop the vegetation more closely, thereby greatly increasing the chances that they will ingest larvae.

Recent research on grazing free-range poultry with livestock has provided preliminary proof that this is an effective and natural way to control parasites. Periodic fecal examinations will monitor any development of resistance. Keep in mind that for any parasite-control program to be effective, all animals kept in the same area must be wormed at the same time.

Proper sanitation is another preventive measure to control parasites. Remove manure frequently and take care to place feed or hay away from con-taminated ground. Fresh, clean water is also important.

Schedules

The age of your animals, the area of confinement, and geography play key roles in worming schedules. In areas infested with meningeal worm (an aberrant type of parasite normally found in deer), you will need to worm as often as every 21 days; where these worms are not a problem, your animals may require only a quarterly schedule. Most nematodes are controlled by injectable ivermectin every 90 days. A nematode problem can be averted by spring worming, before the animals are released to new pasture. Any other livestock species sharing your farm must also be on a regular worming schedule.

Internal Parasites

Some of the most common internal parasites will be discussed here, but these are not the only parasites llamas may encounter.

Fecal Examinations

Your veterinarian can diagnose specific parasites by a fecal examination of individual llamas or from a composite sample collected from a group. The average cost of a fecal exam is $8 to $15. Intestinal parasites usually infect all the animals that are pastured together.

Meningeal Worm

The white-tailed deer is host to the meningeal worm. (A *host* animal is a carrier.) An infected deer passes the worm through its feces. The worm develops into a snail or slug, which the llama ingests through foraging or grazing. The parasite then migrates to the central nervous system, causing neurological problems for the llama, such as paralysis, dragging of limbs, and lack of coordination. The meningeal worm can cause blindness and death. Exposure to this parasite can be kept to a minimum by utilizing adequate fencing. A highly charged electric fence or high-tensile fence would be preferred over wood fencing. In areas of a heavy deer population where there is danger of deer feeding inside the llamas' pasture, it is imperative that your llamas be kept on ivermectin. Check with your veterinarian regarding a schedule.

Eperythrozoon (EPE)

This is a blood-borne parasite that is transmitted through reused syringes, transfusions, or insect bites. The parasite is rare in New England but does occur in other parts of the country. Anemia, high temperature, loss of weight, weakness, inactivity, and lethargy are signs of possible infection. The use of new or sterilized syringes when giving injections will reduce exposure. Keeping llamas away from damp areas that may be mosquito breeding grounds will also help.

Tapeworm

The tapeworm attaches itself to the small intestine of its host. Sometimes pieces of the worm containing eggs are visible to the naked eye when they pass with feces. The life cycle repeats itself when another animal ingests the mite living in the grass. Keeping grazing areas clean and free of manure along with pasture rotation will help to avoid tapeworm.

Coccidia

Coccidia is highly contagious. When an infected animal is diagnosed, remove it from the herd immediately. More common in the fall and winter, it is spread by a carrier that itself may not show signs of infection. In many cases, coccidia is brought in by new animals. This parasite is spread through the feces, ingested by llamas while feeding. In extreme cases, the feces may then contaminate water supplies and/or feeding areas. Maintaining sanitary conditions and being careful of overcrowding will prevent a widespread coccidia outbreak. Fortunately, it is a very treatable disease. Corid added to the animals' drinking water should eliminate the problem; consult your vet.

Mild diarrhea may be a symptom of coccidia; severe cases will show blood in the feces, anemia, weight loss, and straining at the dung pile. Extreme cases left untreated will result in convulsions and death.

Liver Fluke

Liver flukes inhabit marshy areas where snails thrive. The adult liver fluke lays its eggs in the bile ducts of the liver of the host animal (in this case, the llama). The eggs are passed out through the feces and develop in water to a larval stage. The snail, as the intermediate host, carries the free-swimming parasite to vegetation, to which the parasite attaches itself. When ingested by another llama, the parasite penetrates the intestinal wall into the liver, where it remains for about 6 weeks. It then enters the bile ducts until it matures and lays eggs. Avoiding grazing areas that are swampy will decrease liver fluke exposure.

Appetite loss, bleeding from anemia, loss of protein, and digestive problems are the symptoms of liver fluke infection. Permanent liver damage can result if this condition is not diagnosed and treated.

Other parasites such as whipworm, stomach worm, and threadworm are less common internal parasites. Llamas have a naturally high immunity to some parasites found in other species. But occasionally, when housed with other livestock, they may pick up these less common parasites. Keeping your barns, sheds, pens, and grazing areas relatively clean of feces will curtail most internal parasites. Maintaining a regular worming schedule will also help to avoid any substantial problems.

Common Internal Parasites

Parasite	Symptoms	Treatment	How to Control
Meningeal worm	Paralysis; dragging of limbs; lack of coordination	Ivermectin; Panacur	Sanitation; proper fencing to keep out white-tailed deer; regular worming
Eperythrozoon (EPE)	Anemia; high temperature; weight loss; weakness; lethargy	Contact veterinarian (there are several ongoing experimental treatments)	Sterilized or new syringes; reduced grazing in areas where mosquitoes and other biting insects are common; regular worming
Tapeworm	Excessive appetite; visible on feces	Panacur	Sanitation; regular worming
Coccidia	Mild diarrhea; anemia; depression	Corid (as a treatment); Bovatec or Deccox (as a preventive)	Avoid overcrowding of animals; sanitation; regular worming
Liver fluke	Appetite loss; digestive problems; anemia	Ivomec Plus	Avoid grazing in wet and swampy areas; regular worming

Temperature Taking

Using a large animal thermometer for adult llamas and a smaller one for crias, first lubricate with KY Jelly or petroleum jelly. Tie a string on the end of the thermometer and fasten it to a clip or clothespin that you'll attach to the llama's wool — this is your insurance in case the llama defecates and the thermometer slides out. Depending on the manageability of the llama, restraint may be necessary. Normally, if the llama is not overly sensitive on his/her hind end, tying the llama's head should be adequate. With the llama standing, position yourself at the llama's side facing the rear.

After shaking the thermometer down below 95°F, insert the thermometer 1–2½ inches into the anus. Allow at least 3 minutes for the temperature to register. The normal temperature for adult llamas is 99–101.8°F, for babies 100–102.2°F. Call your vet with any questions.

External Parasites

Llamas are susceptible to several types of external parasites as well. A skin scraping or external examination by your veterinarian will determine the specific infection that needs treatment.

Sarcoptic Mange

Highly contagious and the most severe of all manges, sarcoptic mange is caused by a tiny mite that burrows its way into the outer skin layer. It is not species-specific; that is, it can spread from llamas to other animals. Usually attacking the sensitive skin of the legs, chest, belly, or face, this mite causes hair loss, dandruff, and scabs. The llama will itch, and the itching will become more severe as the disease progresses. Eventually, the skin will become thick and crusty and have a leathery appearance. Because this disease is so highly contagious, it should be reported to your state veterinary authority.

Treatment begins with immediate isolation from all other animals. Several ivermectin injections 14 days apart will bring about a fairly rapid recovery. Consult your vet for proper dosages and injection technique.

Requiring a health certificate from a veterinarian for a new llama should be sufficient. However, a new llama may be infected with the parasite in a dormant stage and not yet show any signs of infection. To avoid exposure, isolate new llamas.

Lice

Two types of lice can infest llamas: the sucking louse and the chewing louse. Both are visible to the naked eye. Sucking lice feed exclusively on blood and in extreme cases will cause anemia. Chewing lice feed on debris and hair on the animal's skin surface and will cause skin irritation.

Suspect infestation from lice if you see the animal rubbing or scratching. Other signs are general restlessness, loss of fleece in large areas, unhealthy or ragged fleece, and dandruff. Examine the skin and fiber for the lice or their eggs. (The eggs are about 1mm long and off-white in color. They attach themselves to the hair shafts.) The condition is generally worse in the winter months.

Treat sucking lice with ivermectin. Treat chewing lice with injections of ivermectin and topical carbaryl dust. Not sharing grooming equipment or tack will reduce your chances of spreading lice. Again, if your existing herd is free of lice, a new llama may introduce it into your herd. Isolation for the new llama is best.

Nasal Bots

Nasal bots is a rare condition. It causes a clear nasal discharge in the early stages, which later becomes yellowish and even bloody. Both sheep and deer nasal bots can infect llamas. The adult female fly deposits the immature larvae on the llama's nose. The larvae crawl into the nose and will live in the sinuses and nasal cavities for 2 weeks to 9 months. The larvae irritate these passageways, causing a runny nose, nasal discharge, rubbing of the nose, and extensive sneezing. Eventually, the larvae enter the pupa stage, drop out, and enter the adult stage. The proper dose of injectable ivermectin should be administered by your veterinarian. Providing a shelter or cool place for the llamas, along with fly repellents, will help to avoid exposure.

Ticks

There are numerous species of tick; every region harbors a variety. Ticks feed on blood. They release an anticlotting substance while they feed and in the llama this can cause anemia, loss of appetite, weakness, and infected skin. Some ticks carry diseases — blue tongue and anaplasmosis, for example — which they can pass on through their bite. Anaplasmosis is an infectious bacterial disease found in cattle and caused by the bite of infected insects. It attacks the red blood cells. Blue tongue is a virus caused by biting mosquitoes that has been known to cause abortion in cattle. Neither disease has been found in llamas, and they are rare in livestock overall.

Some ticks release a poison in their saliva. This poison may cause paralysis to the animals, first noticeable by a general lack of coordination, and then particularly affecting the back legs. Breathing will become difficult, and death may ensue. Make a thorough inspection of the skin — possibly requiring some shearing — to help locate one or more ticks. (Ticks will first attack the more tender and thinner skin areas of the ears, legs, under the tail, sheath, or udder, so check there first.) The type of tick largely depends on the geographical area. Keeping llamas out of very tall, grassy areas will lessen any tick problem. In all my years raising and grazing llamas in New England, I have never seen a tick on my llamas. If ticks are a problem, remember that guinea fowl love to dine on ticks; when allowed to free range, they will keep your property virtually tick free.

Ivermectin has been used successfully. An external pour-on dip will treat any overlooked ticks.

Flies

Flies are a fact of life in any livestock operation. They annoy llamas (and people). They require manure on which to lay their eggs. Keeping pasture conditions sanitary by frequent manure removal will help to control flies. Keep your llamas out of ponds and swampy areas to deter other biting insects such as mosquitoes.

Various sprays and wipes are available from feed stores, but their effectiveness depends on the type of fly, breeding grounds, and frequency of application. Horse wipes, sprays, and tags that fasten to halters are other alternatives. Always read the directions carefully when you use any kind of spray. And be sure to cover the llama's eyes.

Ivermectin is broadly used to treat both internal and external parasites. Some parasites have been known to develop a resistance to ivermectin (Ivomec) if used more often than on a quarterly worming schedule. If this is the case, Panacur is a good alternative. There is no way for you to know if your llama has developed this resistance. Ask the seller what the history has been for worming and what wormers had been used. Ivomec is available through most mail-order livestock supply houses. It is available in 50 ml for about $35, 200 ml for $145, or 500 ml for $320. Pour-on Ivomec is available in 250 ml for $37 or 1 liter for $145.

Injectable ivermectin can be administered by you as the llama's owner, but if you have not given livestock injections before, I would recommend you learn under your veterinarian's supervision.

Common External Parasites

Parasite	Symptoms	Treatment	How to Control
Sarcoptic mange	Thick, crusty, leathery skin; excessive scratching	Ivermectin	Isolation of new llamas; regular worming
Lice	Anemia; skin irritation; scratching; loss of fleece	Ivermectin injection; topical carbaryl dust	Isolation of new llamas
Nasal bots	Runny nose; nasal discharge; rubbing of nose; excessive sneezing	Ivermectin	Fly repellents; shelter area for llamas during fly season
Ticks	Labored breathing (extreme infestation); lack of coordination	Injectable or pour-on Ivermectin	Yearly shearing; avoid grazing in overgrown, ungrazed areas
Flies	General annoyance to animals; itching and scratching	Fly wipes and sprays; Ectrin	Fly traps and zappers; sanitation; avoid grazing in wet and swampy areas

Vaccinations

Yearly vaccinations are an important part of herd health care. The vaccinations your animals will need depend on your geographical area. Consult your veterinarian. But a good general rule is to give yearly vaccinations of clostridium C and D, tetanus, and rabies.

Tetanus

These clostridial (oxygen-hating) microorganisms take refuge in an animal's intestinal tract and in the soil. The bacteria multiply in deep puncture wounds, releasing a toxin that affects a llama's nervous system. Spasms of the jaw muscles result (the disease is sometimes called lockjaw). A requirement in your inoculation program, the tetanus vaccine is generally administered in a combination inoculation that includes the vaccine for Enterotoxemia Types C

Recommended Vaccinations

Age	Clostridium C&D **	Tetanus	Rabies	Worming *
Newborn	✓	✓		
Crias at 3 months			✓	✓
6 months				✓
9 months				✓
12 months				✓
Annually	✓	✓	✓	
Pregnant females at 30 days before delivery	✓	✓		

* *Worming schedules vary by region and type of parasite. Worming schedules can range from every 30 days to twice per year. Type of worming products also vary depending on parasite. Quarterly worming is shown here — check with your vet.*
** *Clostridium C&D is a combination inoculation that contains tetanus, enterotoxemia, malignant edema, and blackleg.*
For leptospirosis — Consult with your veterinarian.

and D, as well. Your herd should receive the vaccination yearly and females should get a booster shot 30 days before delivery (be aware, though, that this may cause some stress for the dam).

Enterotoxemia

Also caused by a clostridial organism found in the soil and in the intestinal tract, this is a serious disease that can take a llama's life before any signs of the disease are noticed. Look for llamas that are weak, uncoordinated, and extremely depressed. They may have diarrhea. In severe cases, a llama will have convulsions or lapse into a coma. Overeating, abrupt diet changes, or excessively rich feed can contribute to this disease by causing the bacteria to more rapidly reproduce toxins in the intestines. Yearly inoculation of your entire herd with Enterotoxemia Types C and D vaccine is a must, with a booster to pregnant females 30 days before birthing to maximize the number of antibodies in the colostrum.

Malignant Edema

A third type of clostridial disease is also caused by an organism living in the soil. Malignant edema thrives in the contents of the llama's intestines. Dirty wounds and snakebite are the most common causes of the disease, but check with your veterinarian as to the risk in your area. Swelling around the site of the wound will be severe. While this disease is most common in the midwestern and western portions of the United States, all llama herds should be vaccinated annually.

Blackleg

This clostridial bacteria also resides in the soil and in the intestines of animals. The muscles are attacked when the organism travels from the intestines to the bloodstream. Although it is sometimes difficult to detect, clinical signs are muscle bruising and swelling and, sometimes, fever. The llama can die suddenly, without warning. A yearly inoculation in combination with the vaccine against malignant edema is recommended.

Leptospirosis

An organism that thrives in wet and damp areas such as stagnant or standing water, leptospirosis attacks the kidneys after entering through contaminated food or water or through mucous membranes in the skin. Blood in the urine, diarrhea, loss of appetite, and fever are the clinical signs. Pregnant females may abort their fetuses. Communicable to other livestock, pets, and people, leptospirosis can be prevented by keeping animals out of wet areas and by keeping outside feeding areas clean. Consult with your veterinarian for the type and frequency of vaccination.

Administering Injections

Generally, your veterinarian will give injections. In some situations, a llama must be medicated on a frequent basis for various infections and medical conditions, and its owner will have to learn to give injections.

There are two types of injections: *subcutaneous* (SQ) and *intramuscular* (IM). SQ injections are given between the muscle mass and the skin, usually in the shoulder area, the lower chest down between the front legs, and within about 6 inches behind the elbows. IM injections are given in the muscle mass in the rump, the lower shoulder area just above the elbow, or the thick muscle mass located on the hind end.

Step 1. *After determining the proper dosage, pull up on the plunger until the rubber washer is at the dosage level. Do not remove the plastic cap over the needle.*

Step 2. *Sterilize the rubber top of the bottle of medication with a cotton ball moistened with rubbing alcohol. Give the alcohol a few seconds to dry, then remove the plastic protective cap on the needle and insert the needle through the rubber top. In order to equalize the pressure of the medication you are removing, you must inject air into the bottle.*

Step 3. *Holding the medication bottle upside down without removing the needle, pull back on the plunger to allow the correct dosage to enter the needle. Remove the needle from the bottle and hold the needle vertically. Gently tap it to force any air bubbles to the surface. Expel the air by giving a gentle push on the plunger.*

 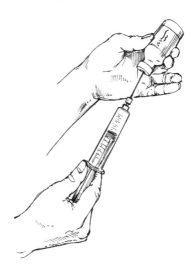

Step 4 *Restrain the llama. Clean the injection site with a cotton ball moistened with alcohol. Stick the needle into the injection site. Before depressing the plunger, gently pull back on it to check that no blood comes back into the needle.* If blood draws into the needle, do not administer. *Withdraw the needle and try a different location.* Never inject medication if blood is first drawn back by aspiration. *(Each new injection site requires cleaning with alcohol and using a new needle.)*

Rabies

Llamas have been diagnosed with rabies. As the incidence of rabies is on the rise — particularly in the East — it is highly recommended that all llama herds be vaccinated yearly with a killed vaccine (a vaccine containing only dead organisms). Rabies is normally contracted from the bite of an infected animal, such as a raccoon. Coyotes, bats, skunks, and bobcats can also be carriers, as can domestic animals. The carrier depends on your geographical area.

Your vet should administer the rabies vaccine, although how much protection it offers is debatable. At present, there is no rabies vaccine on the market approved specifically for llamas.

Hyperthermia and Hypothermia

While llamas adapt well to most climates, they do have some difficulty with extremes of heat and cold. (The temperatures in their natural habitat in South America range from 15° to 60°F.) Heat in combination with high humidity can be fatal to a llama. Depending on your herd management, what type of shelters you have, and if you practice shearing, llamas should be able to comfortably tolerate heat up to 90°F. Shelter from hot sun must be offered along with plenty of fresh water. As for hypothermia, some llamas may start to show signs of it at a temperature of 20°F. Color, age, weight, and other considerations such as pregnancy come into play in extremes of heat and cold.

Heat Stress (Hyperthermia)

With the now-accepted practice of shearing, cases of *hyperthermia* (heat stress) have decreased in the last several years. Heavily wooled llamas are most susceptible to heat stress, particularly those with dense or matted fleece, which impedes air circulation. A llama with a brushed-out coat is less likely to suffer from heat stress than is an animal with a heavily matted coat. Keep in mind, however, that some llamas are simply more susceptible to heat stress than others.

Determining Heat Stress

Color, age, body condition, and weight influence a llama's reaction to heat. Heavily wooled llamas, overweight llamas, first-time dams, newborn crias, and nervous studs are the most vulnerable. A heat-stressed llama may or may not be off his feed, but he will breathe heavily, foam at the mouth, and sway from side to side when standing. If the llama lies down, he is likely to do so in a kush position or even on his side. A llama will have a rising rectal temperature from 103 to 104°F. Males will also exhibit scrotal swelling — this is an easy way to recognize heat stress.

Dealing with Heat Stress

Before shearing gained acceptance, many breeders learned the skill when they were faced with saving the life of a heat-stressed llama. When a llama is down with heat stress, avoid shearing to the skin to prevent shock. Using a pair of hand shears, leave about 2 inches of fleece. (Avoid electric shears; they'll only inflict more stress.)

You must bring the llama's temperature down in a hurry, and a thorough soaking with cool (never cold!) water is the best bet — if you can run a hose from a kitchen sink, you can soak with water in the ideal temperature range of 70–75°F. Hosing and an immediate shearing, most important in the neck axilla inguinal area, the underside, and the legs, will bring the animal's temperature down almost a full degree. It's a good idea to call the vet, too.

If your area is experiencing a humid heat wave, take your llamas' temperatures periodically to monitor how they're faring. Add electrolytes to the animals' water during hot spells to keep your llamas hydrated as excessive sweating can result in dehydration, a serious condition. Electrolytes are ions such as sodium, calcium, and phosphorus that float in the bloodstream. There must be a balance of these ions to maintain normal metabolism. When the natural balance of these ions is interrupted, as in heat stress, electrolytes must be added to drinking water as a supplement. Electrolyte supplements are available from your veterinarian or feed store. Not all llamas are susceptible to heat stress, but it can strike all types and ages of llamas. By getting to know your animals and their habits, you'll know when they are not feeling up to par.

Hypothermia

Prolonged exposure to cold may cause *hypothermia*. Hypothermia begins when the adult llama's body temperature drops to below 90°F. Thin llamas, older llamas, young llamas with less of a hair coat, and llamas with a nutritional deficiency are at added risk. A newborn cria's temperature should not drop below 100°F. Babies that are born outside and not found immediately are susceptible.

Determining Hypothermia

Shivering — the body's way of producing heat through muscle contractions — may be the first sign of hypothermia. When the available energy is depleted, however, shivering ceases and the body temperature begins to drop. Breathing will become labored and blood pressure will fall. Take your llama's temperature if you suspect hypothermia.

Treating Hypothermia

Proper nutrition is the best method of avoiding hypothermia. (See Chapter 3). But if your llama is affected, bring him into a heated area immediately. Keep him there until his temperature rises to normal. If you are unable to move the animal, use an electric blanket (be careful — it can burn), hair dryers, or heat lamps to raise his temperature. Warm water applied with wet towels to the neck, back, sides, and legs of the animal can be used, but don't allow the llama to get a chill.

Crias are highly susceptible to hypothermia. A rectal temperature of less than 92°F (33°C) is a bad sign, and it is unlikely that the cria will be able to rewarm on its own. To bring its temperature up, you should immerse the cria in warm (85°F to 100°F) water, keeping the head up. Check the llama's temperature frequently: You want the cria to warm up steadily but gradually. When the proper rectal temperature has been achieved (100°F to 102.2°F), remove the cria from the warm bath and dry it carefully and thoroughly with a blow dryer.

Eyes and Teeth

Llamas are no more prone to infections and injuries to the eyes and teeth than other livestock are. Regular veterinary care and good herd management will keep any problem from escalating.

Eyes

The eyes of llamas are normally large and brown. Blue-eyed llamas are less common; this trait is believed to be caused by a recessive (hidden) gene. As in other species, some breeders feel that blue eyes are an undesirable trait and should not be bred back into a herd.

Eyes that are tearing, a closed eye, or eyelids that are drooping are some signs that there is an injury or that an infection has set in.

Conjunctivitis

Conjunctivitis is commonly known as pink eye, although conjunctivitis can include various other eye infections or injuries. You may notice a clouding effect, with the eye tearing excessively and the membranes a brighter pink than usual. The llama may also rub his face. To treat, first flush out the eye with a rubber syringe containing saline eyewash to examine the extent of the infection or injury. The llama will probably need to be restrained.

Depending on the llama, full restraint in a chute may be necessary, whereas a more well-behaved llama will accept you working with him while just tied tightly by a lead rope. Remember, llamas are normally head shy, and the difficulty involved in times such as this will depend on whether your llama was imprinted from birth and how much he was worked with.

Dirt and foreign objects may be caught in the eye, and merely flushing it with a sterile saline solution and applying ophthalmic ointment is all that is needed to clear up the problem. Ophthalmic ointment is available from your veterinarian or your pharmacy. Ointments that are used for humans can be used for llama eye problems as well. Be sure it is a new tube of ointment without any previous contamination. Be careful not to allow the tip of the opened tube of ointment to touch anything or contamination will occur. If there is no improvement in a day or two, contact your veterinarian, as there may be a more serious problem, perhaps involving the cornea itself or even the eyelid.

Teeth

Llamas use their teeth for grinding. Any obvious tooth problems should be taken care of promptly, as the animal may be in pain. Abscesses develop on occasion when teeth break during a fight, or in older teeth that become infected. Lumps or bumps in the mouth, foaming, and salivating are indications of possible infection.

"Baby" Teeth

Llamas do lose their "baby" teeth, which are replaced with permanent ones. Occasionally you may notice that your younger llama has a bloody mouth — there is usually little cause for concern. All permanent teeth will normally be in place by age 5.

Older Llamas

As llamas become older, they may lose their incisors. Because it then becomes difficult for them to forage, you'll need to supplement their diet with grain. If they are also losing their grinding teeth (molars), they won't be able to chew the grain, which is necessary for proper digestion. You may actually think your llama has a tapeworm because of the large amount of food he is eating, his aggressiveness toward food, and yet simultaneous weight loss. Moisten the grain in a mash to soften it so he can digest it and receive the proper nutrients. (See Chapter 4 for more about mash.)

Removal of Fighting Teeth in the Male

Fighting teeth start to erupt in the male when he is about 2 years old. All six of these teeth will be present by age 4. These razor-sharp teeth evolved for use in combat between competing males and may have also served in defense against predators. This was Mother Nature's way of selecting the strongest individuals to perpetuate positive traits in the species. (Females may also have fighting teeth, but they serve no purpose that we know of.)

Some breeders choose to leave in these teeth, but removal makes sense if the llamas are housed with others and are overly aggressive. Most breeders will opt to have these fighting teeth sawed off. All my breeding males have had their fighting teeth sawed off, but my gelding packers, even though they have fighting teeth, have not. Normally geldings are not aggressive enough to utilize these teeth on other llamas. But if there is any doubt about any possible fighting, by all means, have these teeth sawed off.

The fighting teeth are in fact not "removed," but are sawed off at the gum line, using, most commonly, an obstetrical wire. While this is a fairly simple procedure, ask for assistance from an experienced breeder or vet if you are uncomfortable performing it for the first time.

1. Restrain the llama's head securely. Two people are needed: one to hold the mouth open and lips away, the other to saw.

2. Form a loop around the base of the tooth. Keep the wire taught and move quickly with one or two pulls parallel to the gum line. This makes a groove in the tooth.

3. Proceed with several more pulls — the tooth will come off quickly. Repeat the procedure for the remaining fighting teeth.

Toenail Care

Care of a llama's toenails may be considered part of grooming, but if toenails are left untrimmed, health problems can result. Lameness can be caused by long toenails. Toenails left unattended can force the toes to grow to the side, which makes walking uncomfortable.

Trimming Your Llama's Toenails

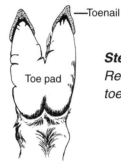

Bottom of foot

Step 1.

Remove debris from toenails and hold one toe apart from the other.

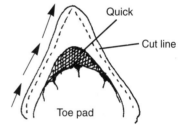

Step 2.

Starting at the rear of one toenail, clip toward the pointed end of the toenail. Repeat on the other side.

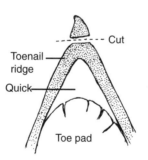

Step 3.

Trim the point of the nail last, being careful not to cut too close to the quick.

Properly trimmed toenails.

If you are a beginner, it is better to trim less at first until you feel confident enough to take more off.

Trimming

Just about all llama owners dread trimming the toenails. Llamas should have their first toenail trimming between 6 months and one year of age. Depending on terrain, nutrition, and individual growth, trimming needs to be done from every 6 to 8 weeks to every 6 months. However, I have some breeding females that only get their toenails trimmed once per year. Llamas that have been desensitized from birth are a joy to work on, but be prepared to securely restrain the average llama. Llamas do *not* like to have their feet handled. Intact males are particularly sensitive. A male whom I imprinted from birth allowed me to touch him anywhere and to pick up his feet for toenail trimming — until he was about 3 years of age, when I started using him for breeding. Ever since he was bred, this male must be restrained for me to work on his feet or legs. Part of an intact male's fighting instinct is to attack the legs of competing males; therefore, any advancement toward a breeding male's feet and legs will be considered an act of aggression.

Toenails that curl to the side will cause the toe to grow irregularly if left untrimmed for long periods of time.

Kevin Kennefick

If a llama's head is adequately restrained, you should be able to work with any part of the animal, including his legs and feet. Llamas tend to lie down when their feet are picked up, so be sure to have them in a chute or tie their heads up.

Overly long toenails can be painful to a llama. The actual trimming of the toenails is not difficult. First remove any debris — mud, manure, and so on — from the nail so you can see where the pad ends and where you need to trim (see Chapter 4, page 81, for a list of the equipment you'll need). Start at the rear of each nail, holding the toe you are working on apart from the adjoining toe to allow room to maneuver the clippers.

The Stomach

Llamas have a three-chambered stomach that is highly efficient at digesting food. On occasion, they do develop some digestive problems, the most common of which are constipation and diarrhea. (Normally, a llama's dung is pelletized and ranges from dark green to brown or black.)

Impaction

A clogged intestine results in an impaction, which leads to colic. Call in your veterinarian. Sometimes the problem is the result of a parasitic infestation; a regular worming program will prevent impactions caused by parasites. Warm water administered in an enema or given orally by syringe is the usual treatment to loosen the impaction. About one ounce of warm water for llamas under one year of age, and at least one pint of warm water for llamas one year and older, will suffice. A rubber ball syringe, available at your local pharmacy, will work fine. *Caution:* Mineral oil aspirated into the lungs can cause pneumonia. This is best done by your vet, or at least under his or her supervision.

Constipation

Constipation results from a change in feeding habits or stress on the part of the llama. Frequent straining at the dung pile is a sign of constipation. Laxatives and enemas can help the situation, but the cause itself must be treated. Changes in feeding habits, such as a new type of feed or an extreme change in forage, may cause constipation. Stresses, such as extended travel, can also result in constipation. A high fiber diet (see Chapter 3 for more information) will prevent any digestive or constipation problems. And as with an impaction, warm water should be administered orally while the animal is restrained in either a chute or by the head. (Female llamas about to give birth may make frequent trips to the dung pile. Do not confuse this with constipation.)

Diarrhea

Diarrhea occurs when liquids are not absorbed through the intestines. Although soft feces are common in newborns, severe diarrhea can be a serious problem (in crias as well as older llamas) that leads to dehydration. *If a cria has diarrhea, is displaying abnormal behavior, and is running a fever, it should receive immediate veterinary attention.* Diarrhea can occur if the cria does not receive colostrum. In older llamas, diarrhea can be caused by various intestinal parasites.

Dehydration is potentially life-threatening. In cold climates, llamas and other livestock become dehydrated when their drinking water is frozen or inaccessible. Signs of dehydration include decreased urination and skin that feels less elastic than usual.

Health Maintenance Record

Name _____ ILR# _____ Blood Type Case # _____

DOB _____ Microchip # _____ M_____ F_____

Castration date (if male) _____ Fighting teeth removal date _____

Vaccinations			Worming			Weight		Toenails Trimmed (dates)	Shorn (dates)	Date Bred	Cria Delivery
Date	Product	Dosage	Date	Product	Dosage	Date	Pounds				

Miscellaneous Comments: _____

_____

_____

_____

_____

_____

_____

First, rehydrate your llama as quickly as possible. Then your vet will want to determine the cause of the diarrhea in order to treat it. Rehydration should be done by your veterinarian. Electrolytes may be given orally, or in extreme cases, intravenously. After the llama is properly rehydrated, you may treat the diarrhea with Kaopectate orally: In crias, ½ ounce; llamas from 6 months to one year, 1 ounce; and llamas one year and older, 2 ounces.

Euthanasia

Whether to bring an animal's life to an end is always a difficult decision, but it's an important topic that must be addressed. Euthanasia may be the kindest option for any animal that is suffering from a serious injury or illness, especially if there is no hope of recovery; for animals that have become severely debilitated due to age; or for those that are extremely aggressive and a danger to the humans around them. Your veterinarian can help you make the decision and provide information about burial, removal, or other alternatives when a llama is euthanized (or dies of natural causes). You can obtain more information on euthanasia from a brochure developed by the American Veterinary Medical Association; your veterinarian should have copies available.

Six

THE HERD SIRE

The herd sire is your foundation if you have a breeding program. Because he will represent your herd, he should be well conditioned, structurally correct, genetically sound, and manageable.

Buying the Herd Sire

An inexperienced owner may purchase a young intact male to begin her program. Unfortunately, most of these "first-time" males are not stud material. Breeders who want to "unload" these lesser males do not always advise the purchaser that they are not of breeding quality. So it's possible some overzealous or uneducated new owners allow these males to make their way into the llama breeding population. However, responsible breeders always try to improve offspring by using the best possible male.

Educate Yourself

Do your homework and visit farms with respected studs before you start shopping for a herd sire. Even before you begin your search:

1. Have a concise and clear business plan.
2. Set goals for your breeding program.
3. Understand conformation and soundness.
4. Know the weaknesses and strengths of your female herd.
5. Consider what would constitute your "ideal" llama.

If you feel that you are not adept enough to evaluate llamas with these considerations in mind, consult an expert. Purchase your sire from an established and reputable breeder. Conformation, soundness, and pedigree are top priorities — after all, the stud you choose will pass on his strengths and weaknesses to every cria born to your herd. Also look at his size or height, bone density (thickness of bone), type of head, and ears. A poor decision could lead to a major setback for your breeding program. Don't be afraid to ask for help!

To develop the ability to select the proper stud for your breeding program, visit as many farms as possible. Attend shows and sales. Inspect a number of animals. When you visit the farm of a potential herd sire, ask to see his sire, dam, grandsire, and granddam, as well as any offspring he may have on the ground. As you examine these branches of his family tree, look for the characteristics that are being passed along. Evaluate how he fits in with your goals.

Budget

We should all have realistic budgets. You may find the perfect herd sire, but he may be out of reach financially. Shop around. Compromise. Consider a full brother to the male you originally wanted, even though he may not be the color you had hoped for. An older male may also fit the bill without the sacrifice of your main objectives.

Reproductive Fitness

Genetic traits are passed on through both the dam and the sire. Research the potential stud's sire and dam, and the grandsire and granddam as well. If you can, find out about the maternal sire, too. Research into the stud's sire and grandsire should show testicles that are average or above-average in size. Both testicles should be visible and oval in shape. Males should have a strong libido, no genetic defects in offspring, and the ability to settle females quickly. Dam and granddam should have the ability to conceive quickly (within second breeding), easy births, no genetic defects in offspring, and good milk production. Both sides should demonstrate prepotency (the ability of the male or female to duplicate himself or herself as closely as possible). Determining prepotency in both parents is as simple as viewing offspring of the sire and dam. Be sure to view their offspring when bred to other females and males as well as to each other. This will give an accurate idea of what the sire and dam tend to throw (produce in offspring).

Ask about the breeding style of the male that interests you:

1. What type of breeder is he?
2. Is he aggressive, but not too aggressive, with maiden females?
3. How long is each breeding he performs?
4. Is he susceptible to heat stress?
5. Does he fight with other males?
6. Is he respectful of fencing?

Color and Fiber

Color is difficult to breed into a herd. If a certain color is one of your goals, engage the services of a consultant who is familiar with genetic combinations and probabilities. Consultants can be found through most llama publications or your veterinarian. The services of a consultant can range from $100 to $500. Some consultants charge a fee based on a percentage of the purchase price of the llama, and some work on an hourly fee plus expenses. Be sure to check the consultant's background, record, and creditability. Ask for references from people who used his/her services. Color preferences are usually a matter of taste; various colors are trendy for a while and then change. Llamas come in as many colors as horses. Many different shades of brown exist, ranging from light beige to dark chocolate. True black (no brown tones) in llamas is rare. Other colors include shades of whites as well as rose colors and beiges, reds, and cinnamons. Gray was once rare, but seems to be appearing more and more through conscious efforts in breeding programs. Llamas also come in mixed colors, black and white, brown and white, and variations and combinations of all the colors listed here. Like horses, llamas can also carry an appaloosa color (white/beige with darker spots all over the body). In the show ring, a judge may choose a darker llama over a conformationally comparable lighter-colored llama, but it still remains a matter of individual judge preference.

If you are breeding animals for fiber, you must be able to distinguish quality fleece. Learn all you can about bloodlines that are known for their fleece production. Figure out what you'll need to improve the quality of fleece in your breeding program. If you are interested in fiber only for its aesthetic value, find a stud with odds in his favor that he will pass on his fiber type.

Genetic Defects

Recessive (hidden) genes may be responsible for certain defects. Carriers of recessive genes often appear as phenotypically normal animals. Animals that have produced offspring with genetic defects may still be out there producing rather than being culled. Remember, the sire and dam can both contribute to a defect, but it takes only one to pass that defect on to 50 percent of its offspring.

Bloodlines

A superior bloodline should consistently produce offspring of high quality. When researching a llama's bloodline, look for evidence in both the stud's sire and his dam's side.

If you are considering the purchase of an import as your herd sire, take into account the difficulty of researching the bloodlines. Some breeders purchase blindly when the main consideration is new blood and an expanded gene pool. That expanded gene pool could breed back into North American herds those undesirable traits that many llama owners have selectively bred out of their herds. It is impossible to detect recessive genetic faults, crossbreeding, linebreeding, or inbreeding (see Chapter 9) when you cannot go back to research previous generations. The purchase of an import could be a wonderful decision, but it's just as likely to have genetically negative results. If you want to consider an import bloodline, choose one that has been in this country for several generations and shows consistent quality in its offspring.

Gale Birutta

North American sires such as this Catskill male are relatively simple to research. These llamas were the very first to be registered with the International Lama Registry. The registry will research any llama for you for a small fee.

Guidelines for Purchasing a Herd Sire

- ◆ Above-average bone density (I always prefer a llama with heavier bone density rather than lighter)
- ◆ Banana ears
- ◆ Excellent conformation (conformation types will vary depending on individual preference)
- ◆ Bloodline (pedigree) traceable to at least three generations, preferably five
- ◆ Well-known sire and dam (I find this to be helpful)
- ◆ Bloodline must demonstrate consistency in prepotency on sire's, dam's, grandsire's, and granddam's sides
- ◆ Proper jaw conformation
- ◆ No genetic defects or faults in bloodlines
- ◆ NO LESS than 3 years of age

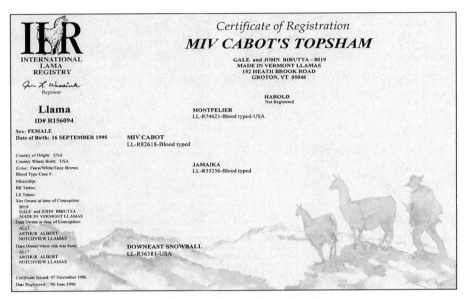

A sample certificate of registration.

Registration

The International Lama Registry is now closed. Presently, it is the only registry for llamas. Make sure the sire is registered, along with his sire and dam. An offspring can only be registered if both its sire and dam are previously registered (or registerable). Offspring of an unregistered sire (or dam) is less marketable, and anyone looking to purchase breeding stock is unlikely to consider it.

Once your herd sire has joined your farm, it will be helpful to be able to document his heritage and explain its benefits to potential buyers of his stud service or of his offspring. It is common for breeders to promote their sire and offer outside breedings — and then cease offering outside breedings. Once a particular sire becomes recognized as producing top-quality offspring, the breeder has created a supply and demand situation and closes outside breedings. In this case, if you want an offspring from that particular sire, you must purchase one from the sire owner. I have done this with my top two sires, Montpelier and MIV Cabot. I have created a supply and demand situation, and offspring by these two sires are sold before they are even born.

Managing the Herd Sire

Management of the herd sire should begin when he is young. Young males should be separated from females by the age of one year. Usually they are not physically developed enough to breed, but it has happened. For the most part, males can physically inseminate a female at the age of 1½ to 2 years. Two years old is a good target for breeding your young male. If you have produced your own potential sire, make sure you aren't "barn blind." Evaluate him as you would any other animal. Other than his genetic background, the main concerns are that he produces, stays healthy, and remains stress-free.

Preparing for Breeding

The stud works hard during breeding season, particularly if he has many females to service. A fit and potent herd sire can breed as many as 15 females per week. This varies depending on the age; an older sire (12 years and up) may be only able to breed several times per week. He may become nervous with anticipation. To reduce stress and maintain his ideal weight, keep him on a good, nutritionally balanced diet. *Slightly* overconditioned (heavier than normal) is best for the sire at breeding time. Extra weight, in moderation, will keep his energy level up, his interest high, and his performance at its best.

Before any breeding takes place, check the sire's toenails. They should be properly trimmed to avoid injuring the female. Check his legs for any cuts or injuries so that breeding is not painful or unpleasant for him.

Male Reproductive Anatomy

Two testicles in a scrotal sac should be apparent just below the anus. The sac holding the penis, called the *prepuce,* normally points toward the tail. When the penis is erect, the prepuce points forward and the penis extends from it. Males reach sexual maturity and are capable of copulation when they are 2 or 3 years old; sometimes they are *capable* of copulation much earlier but they are not yet physically mature.

Males, even at an early age, "practice" by mounting their moms, female pasture mates, and even other males. In a youngster, the prepuce and penis are connected by tissue, making extension of the penis impossible. As the young male approaches sexual maturity, this connective tissue disappears, allowing the penis to extend during sexual excitement or copulation.

The Inexperienced/Young Sire

Some inexperienced males may be put off by a single spit from a dominant female. Others become so excited or impatient that they forget which end is up. With a little help, a young stud can be guided to the proper end and helped to get into the proper position. With experience, young males will quickly become more dominant. By the time they are 3 or 4 years old, they'll be fully capable.

Try not to breed an inexperienced male to a maiden female. Use a reliable, easily bred, docile female who will not intimidate him on his "first time out." Keep the female's head pointed away from him lest she decide to spit. It is a good idea to wrap the female's tail with an ace bandage to assist the inexperienced male in accomplishing his objective and to prevent introducing hair into the vagina. (See Chapter 9, to learn how to wrap a female's tail.)

The Crushed Ego

In the summer of 1996 we sold MIV Sheffield to good friends of ours, Doreen and Curt Hayes of Windchill Hill Llamas, New York.

Doreen Hayes

MIV Sheffield was sold as a herd sire but had not yet been bred. After he had settled in for a few days, Doreen and Curt brought Sheffield to their only adult female. Apparently she was already bred; she spit and carried on and thoroughly rejected him. Young males who have not been bred before and experience this kind of refusal go off in a corner and pout. And this is exactly what Sheffield did.

Melody Violet Hayes and her favorite llama, MIV Sheffield.

About a month later, Doreen and Curt purchased additional females that were due to be bred. Curt put Sheffield in with the new girls, but Sheffield did nothing. In fact, he ran the other way. Again, this is quite normal for well-mannered young males who have initially been rejected by a female. New to the breeding business, a distraught Curt phoned several times to find out what was wrong. I told him to give Sheffield some time, that he would come around on his own. I will never forget what Curt said: "It's very frustrating; it's like having a new Christmas present without the batteries."

MIV Sheffield has since gained confidence, established his territory, and now takes care of business with enthusiasm.

Male Refusal

Occasionally you will notice a male experiencing problems settling in to breed the female. Sometimes the problem is with the female; she could be bred already or have some physical problem (see Chapter 7). Learn to read your stud's behavior; you'll gain insight into some breeding problems and it may help you to determine pregnancy in females he refuses to breed.

The male is most comfortable in his own pasture or paddock. Keep all other females out of his sight so he will concentrate on the female at hand. A few males like females who play hard to get. These males may not attempt to breed unless the females put up a fight or run away.

Male Infertility

Infertility in adult male llamas is uncommon, but there are a number of factors that can render a male either sterile or infertile. Sterility is permanent; infertility may be temporary. Either condition can have an anatomical, metabolic, or behavioral cause, or the llama may have an infection.

If you suspect an anatomical problem, your vet may have to sedate him to perform an accurate examination. He should be checked to determine if he has all the proper parts and that their location, size, and shape are correct.

Hormone deficiencies can cause problems with sperm count and formation and/or testosterone levels after an animal reaches maturity. Some illnesses related to poor nutrition may cause metabolic infertility affecting sperm production and quality (see Chapter 3). The most common metabolic problem is heat stress. In warmer climates, many breeders simply accept the fact that their male may be sterile during the hotter months. This temporary condition corrects itself with the onset of cooler weather. Other causes of infertility include nutritional imbalance, mineral deficiencies, and obesity. Check with your vet for an accurate diagnosis.

Some males become overaggressive or so excited when they see a female that they get confused. They then cannot settle down and breed. If there are other adult males around, some males may perceive them as competition. The priority then becomes protection of territory. Other behavioral problems arise when a breeder interferes too much with a young, timid male or when he has been traumatized by an older, overly aggressive female. Sometimes the problem is just plain exhaustion.

If the penis is injured during breeding or in fighting with other males, serious problems can result. Checking for infections in this area should be routine.

Checking for a penile infection is best done by your veterinarian if you are a novice. The llama will probably have to be restrained or sedated. The veterinarian can palpate (feel with hands) the penis forward and check for signs of infection. In breeding llamas with long or dense wool, the tip of the penis may become entangled with wool, causing an infection after being drawn back into the prepuce. (This is why it's a good idea to wrap the female's tail during breeding.)

Your llama may present other problems with breeding. Consult your vet for advice.

If you and your vet have ruled out all other problems, try a semen evaluation. Several factors are considered when evaluating semen: the age of the male, his general health and condition, and how much and how recently he has been bred. Several samples should be taken on different occasions to get an accurate sperm count. Semen samples can be retrieved from the vagina of a female that he has just bred. Again, consult your vet.

Factors Affecting Male Potency

- ◆ Anatomical problems
- ◆ Hormone deficiency
- ◆ Illness/infections
- ◆ Heat stress
- ◆ Nutritional imbalance
- ◆ Obesity
- ◆ Overaggressiveness/excitement
- ◆ Excessive interference by the breeder
- ◆ Exhaustion
- ◆ Timidity

Showcasing the Herd Sire

Promotion of a herd sire is the lifeblood of many farms and ranches. If you wish to offer stud services, it is important to market your herd sire when and where the opportunities arise.

Choosing to Promote

A breeder must first decide whether it is in her best interest to promote her herd sire. Knowing why is also important. Even though the industry continues to change and grow, no type standard has yet been established for llamas. This is important knowledge for the new owner or breeder and for the small to mid-sized enterprise. The selection of a certain "type" of sire in the beginning may be a disappointment later after you spend time, money, and effort in promoting a herd sire that has failed to meet the current criteria. Make sure he has promotable credentials, that the majority of his offspring are of breeding quality, and that he and/or his offspring will place well at shows.

The Home Front

A potential buyer's first impressions of your farm are important. Stable your herd sire where he will be in a prominent position. You want a buyer to be struck immediately by the sight of the most magnificent creature he has ever seen — your stud! An area close to an entrance, perhaps along a driveway, will showcase your sire. Male llamas are proud of their territory, and enjoy presenting it to visitors.

An occasional thorough grooming is important to keep him looking his best. Frequent surface groomings help to maintain a tidy appearance. See more on grooming in Chapter 13.

A herd sire standing proudly at the entrance to your farm will impress potential buyers.

Grooming

Surface Grooming. Brushing the top coat only to remove hay, straw, or debris on outer coat; this is a good, quick grooming for packers. This should take not more than a few minutes each day.

Thorough Grooming. Brushing to the skin and removing all mats while separating all fibers; this should be reserved for shows and exhibitions. Depending on how often you thoroughly groom your llama, this can take anywhere from one hour to several hours. Utilization of coat conditioners and a Circuiteer blower is imperative.

Differences. Mats and debris in undercoat will remain in fleece with surface grooming, whereas all mats and debris will be removed in thorough grooming.

The stud service buyer will likely want to inspect the rest of your facilities. He will want to know that you manage your business and herd well, and he will want to feel comfortable about leaving his female with you for breeding.

Open houses and farm tours are excellent ways to promote your farm and stud. A tremendous amount of work is required, but it will pay off in the number of potential customers who will visit your farm and view your sire and his offspring. They'll remember your farm when they're in the market for a stud or new llama.

Showing Your Sire

Showing your herd sire can either catapult him into the limelight or "bury your farm."

It is important to have faith in your own breeding program. If you feel you should not show for fear of losing, then you lack confidence in your own breeding program and should take a closer look at your stock and goals. Showing provides an excellent opportunity to evaluate your program and to compare your stock to that of other breeders. The judges make their decision based on what they see at that show on that day. If your sire is clearly the best at the show but is having a bad day or misbehaving, you will likely be placed accordingly. Only one sire can win, so be prepared to be a gracious winner or a sportsmanlike nonwinner. Keep in mind, as both a breeder and an exhibitor, that you should not let any one show cloud your judgment regarding what you feel is the correct sire for your breeding program.

You will take a risk if you decide to show your herd sire. There are many potential customers observing the outcome of each show, and they may base their decisions on the judge's awards. If your sire shows poorly, you may lose these customers.

If you do decide to exhibit your male and he becomes a

Photos by Ron

Capturing Grand Champion consistently at shows will make a star of your stud.

consistent winner, then by all means, show your sire. Llamas that consistently win lend considerable benefit to your breeding program. Winning Grand Championship after Grand Championship year after year will send your herd sire to the very top. It is not difficult for an animal to win a championship title at some point, but when that sire continues to win everywhere he goes, season after season, the llama world tends to pay attention. But please remember, you don't have to show your sire. Many top sires have never set foot in a show ring but have been promoted through advertising and word of mouth.

The author's MIV Cabot, winner of six championship placings, remains one of the top show sires in the Northeast.

Critique Your Sire

If you want to show your herd sire, stand back and take a serious look at him. Be objective. Be honest and critical. Ask a friend to handle him for you while you take a good, long, "judge's" look. It may also be beneficial to get an outside opinion. Check for the following:

- Is he tracking correctly, both toward and away from you? Are all his toes pointed forward and not angling away from him? At a brisk walk, he should be moving on a straight line and with a smooth gait and good balance.
- Is his topline strong and straight? Ask your friend to "square him up" (to set him up to stand for the judge). Critique his legs. Are they straight or are there visible faults?
- Does he have an attractive head with good, well-set ears and no jawline under- or overbite?
- Does his neck have good length-to-body ratio? His neck and legs should be in proportion to the rest of his body. The neck should

not be too long or short and the leg length should be balanced to the body. (This may be difficult to determine with a heavily wooled sire.)

◆ Does he possess that special presence that allows him to stand proudly in the show ring and strut his stuff? His attitude should say that he is the best.

Squaring or Setting Up

Squaring, or setting up, takes some practice. *Squaring* simply means placing your animal so that he stands collected and squarely on all four feet. He should not be leaning backward or forward.

All four feet should be evenly spaced and neither too close nor too far apart. To set up your llama, turn his head to the left — the right hind leg will go back.

This llama has been squared up to his best advantage: His legs are properly set and his head is held high.

Set up the left hind leg by turning your llama's head to the right. These maneuvers place the llama temporarily off balance, forcing him to place his feet properly for even weight distribution.

When correcting your llama's set up, do not pull forward and down on the lead. This is a natural tendency but it will put the llama off balance, producing an uneven distribution of weight. When you pull forward, the llama will put more weight on his front legs, causing his neck to go down and the topline to round. By pulling up and back, the llama's head will go up, and he will transfer his front weight onto his back legs and straighten his topline.

Attitude

Attitude — his — is a consideration in the decision to show your sire. He must be manageable and willing, able to settle in to a new environment, and be a "gentleman." He should lead easily in the presence of open females. He should allow the judge to touch him for hands-on evaluation. He must be able to stand quietly and obediently for long periods of time in large classes. (See Chapter 11 for more on showing and exhibiting.)

Performance

Although training for performance showing takes considerable time, llama buyers are often more impressed with the "working" herd sire rather than just another pretty face. Performance work shows that he is intelligent, willing, and manageable, has an even temperament, and is structurally able to work. Because these are desirable traits that he can pass on to his offspring, it's a good idea to demonstrate them.

The Herd Sire Portfolio

The herd sire portfolio is an important tool if you want to offer outside breedings. The portfolio should be professionally packaged and should include the following:

- 8 x 10 color photos of the sire.
- Genealogy; a copy of the ILR certificate. Include any additional information on the sire's pedigree, his dam and sire, and his granddam and grandsire. Describe their colors and sizes, too.
- Photographs of his offspring to exhibit prepotency, color, fiber quality, conformation, and hereditary traits such as phenotype.
- Reproductive history, including any problems on the maternal side of the pedigree, and hereditary defects.
- Show records of your sire, his offspring, and his siblings.

The herd sire portfolio may also include a copy of your standard breeding contract and a description of your benefits package.

Offspring

If you show the offspring and they do well, you have reached the next level in promoting your sire through shows. The time has come to retire your sire from the show ring and to promote him instead through the successes of his offspring.

Farm Displays

If the show can accommodate farm displays, design an eye-catching exhibit that promotes your farm and your sire. Create a herd sire portfolio with color photos of his offspring, sire, and dam. All pertinent information, such as show wins and his pedigree, should be included. (See Chapter 11 for designing a farm display.)

Gale Birutta

A herd sire's offspring are the proof that he is producing what he was meant to.

Advertising

Often, advertising revolves around a sire and his offspring. These ads create a perception of demand in the market as well as quality. Other farms may decide to bring your sire's bloodlines into their own breeding program with the hope of improving their herd and adding a popular herd sire's name to their offsprings' pedigree — furthering their own marketing prospects. The decision to purchase breeding stock or outside breedings may be influenced by advertising and the reputation of a particular herd sire.

A small but consistent advertising campaign is better than a splashy, one-time-only ad. Contact magazines and ask for advertising rates and ad packages. Some publications to consider are your regional llama association's newsletter (see Appendix A), state llama organizations (see Appendix A), and various other llama magazines and publications listed in Appendix C. Rates for ads in local and regional publications can run as low as $5 for a business card–sized ad to as high as $2,000 for a full-page color ad.

Include photographs in your ads. They must be of high quality to show your animals to their best advantage. Make sure your llamas are well groomed

Quality photographs and catchy wording highlight this sample advertisement for regional llama publications.

for the photo session. Some pictures should feature the sire. Photos of his progeny may result in the sale of some or all of the offspring. And pay attention to objects in the photos that might distract from your animals.

Take advantage of recent show wins when designing your ads. If a relative of your sire sells for a record amount, or shows well, by all means mention this in your ad. It supports your claim that your sire and his lineage are valuable and that his genes produce consistent quality.

Advertising should be professional looking and truthful. It needn't be expensive. To get ideas, study ads that other breeders are running. But be careful not to copy or get too close to the promotional techniques of another. Even if you think you're not creative, it won't be as difficult as you think to come up with some promotional ideas for your herd sire. Often a magazine or newspaper's ad representative will help you to write copy and design the layout.

Breeding Services

Caring for outside females is a tremendous responsibility. You must have adequate facilities and the time to devote to breeding and caring for these visitors. Be prepared to treat these females better than you do your own. Be ready to deal with problem females. Preparation for problem females should start with specifics in writing in the breeding contract. If the female has had previous conception problems, then spell out in the contract what you will and will not offer. Don't offer open-ended free board while the female is being bred, as you may have her for 6 months if she is a difficult breeder. Discuss and put in

MADE IN VERMONT LLAMAS
BREEDING SERVICE AGREEMENT

AGREEMENT of breeding service(s) made this _____ day of
_____, 1996, by and between the parties identified in Paragraph 1
below. _____ agrees to breeding
services from MADE IN VERMONT LLAMAS on the terms and conditions of this Agreement hereinafter set forth below:

1. PARTIES:
The party referred to in this Agreement as the BREEDER is:

MADE IN VERMONT LLAMAS
152 Heath Brook Road
Groton, VT 05046

The party referred to in this Agreement as the PURCHASER is:

_____
_____
_____
_____

2. DESCRIPTION OF LLAMA FOR SERVICE:

MIV CABOT
DOB 10/16/90, ILR# 82618
BLOOD TYPE CASE NO.: LL3882-2

3. PAYMENT: Breeding fee to MIV CABOT is $2,000.00 with a
25% deposit of $500.00 due by _____, 1995. The additional 25%
($500.00)is due upon delivery of female(s) for breeding, with the bal-
ance of 50% ($1,000.00) due at time of pick up of pregnant female. It is
the responsibility of _____ to deliver
said female(s) for breeding to MADE IN VERMONT LLAMAS, as well as
picking up bred female(s).

4. It is the financial responsibility of _____
_____for the proper veterinary certificates
for interstate travel both to and from Vermont for female llama(s).
Breeding fees include full unlimited board for up to 60 days with an
inclusive internal ultrasound test to determine pregnancy.

5. This Agreement includes a live birth guarantee, but does not
designate guaranteed male or female cria. "Live birth" is is deemed a
normal,healthy cria that has stood and nursed.

6. Special services provided by MADE IN VERMONT LLAMAS
shall include:
   a. conditioning program for overweight female;
   b. cria by dam's side shall be halter trained to lead;
   c. dam shall be shorn and show groomed before departure

7. This Agreement constitutes the entire agreement between the
named parties. It is further agreed that any modifications to this agreement be executed in writing as an addendum.

DATED _____

DATED _____   Gale or John Birutta
                       MADE IN VERMONT LLAMAS

Sample breeding contract.

writing that the owner of the female will be responsible for all costs in dealing
with a female with conception problems. Veterinary fees and other costs associated with fertility specialists or hormone injections should be the financial
responsibility of the dam's owner. Some owners believe that the male will
solve any conception problems their female may have. If the female has not
settled previously with another proven male, then it is likely she has a problem. Be leery of accepting her for breeding.

Defining Your Customers' Needs

The breeder needs to understand the goals and expectations of the customers. It is important that the herd sire they have chosen (yours) will meet
their expectations and produce the type of offspring they expect. Have them
define for you the strengths and weaknesses of the particular female and how
they feel your herd sire will complement her.

It makes good business sense to accept any female for breeding, provided
she is sound, conformationally correct, and has no history of genetic defects in
her offspring. Fancy phenotype appearance should have absolutely no bearing.
Even the most beautifully conformed female may not throw desirable offspring. There is no better way to promote your stud than by allowing him to
upgrade an average female's offspring.

Written Contracts

The key to a clear understanding of what is expected on both sides is a written contract. Your contract should include what the female's owner expects from the stud; the services the stud's owner provides; terms and conditions of the contract; and what the sire's owner is able to do to ensure the breeder's satisfaction. The goal is communication without misunderstandings.

Many contracts require up to a 50 percent deposit at the time the contract is signed, with up to 25 percent due when the female arrives at your farm and the balance due upon confirmation of pregnancy before she returns to her owner. The owner of the sire should include the cost of pregnancy determination in the stud fees. A veterinarian can perform an ultrasound and a blood progesterone test for pregnancy determination.

The Edge

There are many breeders offering services to exceptionally fine studs, so ask yourself what *your* farm or stud can offer that other breeders cannot. If possible, offer services that will set you apart.

If you have shown your sire successfully, establish your breeding fee to accommodate the recent show recognition — then add special services.

- ♦ Knowing that many breeding-age females are overweight, offer a special diet and exercise to bring the female into "satisfactory birthing condition."
- ♦ If you have show grooming experience, offer a complete show grooming for the female before she leaves your care.
- ♦ Offer a complimentary shearing, especially if the female is with you in the spring or summer.
- ♦ Does she come with a cria by her side? Offer halter breaking and lead training.

Include all these "extras" in the base breeding fee. Other extras could include free transportation for the female, the upgrade of an internal ultrasound rather than a progesterone test for pregnancy confirmation, free rebreedings if a cria is not satisfactory, or multiple breeding discounts. These services may influence a customer's decision about which stud and breeder to use. If you offer extra services, make sure you get the owner's permission, in writing, before any are rendered.

Risks for Your Stud

Risks for your stud are reduced when the service is provided on your farm, but they still do exist. Request a health examination of the female to screen for infections. The female's owner is responsible for providing a clean bill of health. If the female is woolly, she should be clipped around her perineal area (genital area) by her owner (or permission granted to the sire's owner to do so) to prevent any interference from wool during breeding. Tail wrapping at the time of breeding is also a good idea, and is usually done by the sire's owner just prior to introducing the male to the female.

Leasing Out Your Stud

You may prefer not to let your finest llama leave your farm, but as the owner of the sire you can benefit immensely. The practice of leasing out a stud gives breeders in the host farm's region a chance to see your magnificent sire when they otherwise may not. This is good promotion for your enterprise. It also endorses your sire because the host farm has chosen to breed a number of his females to him. A demand is created that can bring additional outside breeding opportunities for your farm.

Know the host farm's experience and be sure to inspect the facilities. Spend some time with your sire's hosts to familiarize them with your sire's breeding techniques and attitudes. They will have to learn how to "read" your sire and how to treat and handle him.

If you do not carry full mortality insurance on your sire, consider it now. Make sure you value your stud realistically and include any replacement costs in the amount you do elect to purchase. This assessment will probably require a detailed examination by a veterinarian.

SEVEN

THE DAM

The dam is half of any breeding program and should not be viewed as secondary to the herd sire. A high-quality prepotent male may upgrade his offspring, as may high-quality females who are prepotent. Prepotency is the ability of the male or female to duplicate himself or herself as closely as possible. Offspring will reflect the genetics of both parents.

Gale Birutta

This female has duplicated herself. The similarity in her offspring is remarkable.

Breeding Age

The optimum age to breed a maiden female depends on the management practices of the breeder and the genetic predisposition of the female. It is important to separate breeding-age males from female crias by weaning time. Early conception can occur in some fertile young females. When a female becomes pregnant at too young an age, she is at risk for having a difficult or premature birth, producing a low birthweight cria, and having a more difficult time bonding with her baby.

On the average, try to wait until the maiden female is at least 18 months of age. If she is lightweight or unusually small, give her an extra six months before breeding her. Sexual maturity is also tied to the female's mental attitude. She may be 2 or 3 years old, but if she isn't mentally ready, then she may not conceive. If she is mentally ready, she'll readily tease the male. Many breeders spend big dollars in veterinary bills to find out why their female is not yet bred although her body is ready. It is not uncommon to see females over the age of 5 pregnant for their first time.

> Female alpacas mature at approximately 14 months of age. Birthing and delivery are the same as in llamas.

Female Reproductive Anatomy

The vaginal opening of the female, the vulva, is 1–1½ inches long. Proceeding through the vulva is the vaginal opening, which leads to the cervix and then to the uterus.

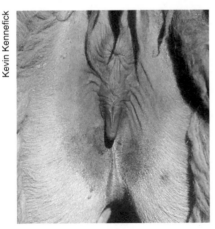

External female anatomy.

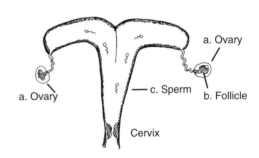

Ovaries and uterus of a recently ovulated female: a. ovary; b. follicle; c. provided the male is fertile, spermatozoa will enter the uterus.

Kevin Kennefick

The Reproductive Cycle

To evaluate any reproductive problems, you must understand the reproductive cycle. Unlike most other livestock, llamas are induced ovulators (rabbits and cats are, too). This means that the female llama will enter her heat when introduced to and stimulated by the male. Llamas do not have regular "heat" cycles. After stimulation by the male and penetration of the cervix, the female should ovulate. Her ovaries should produce a new follicle or egg approximately every 10 days.

Because the female does not display physical signs of receptivity, the male llama depends on behavioral signals. Most open (nonpregnant) females will tease the male by pacing the fence line or by lying down in front of him. (Unfortunately, some females will also spit, run, and even charge the male when open.)

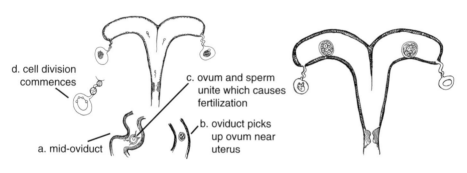

d. cell division commences

c. ovum and sperm unite which causes fertilization

b. oviduct picks up ovum near uterus

a. mid-oviduct

a. Mid-oviduct. b. The oviduct will pick up the ovum near the uterus.
c. Ovum and sperm unite, which causes fertilization. d. Cell division commences.

Uterus of a recently impregnated llama.

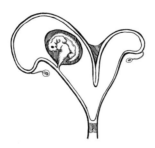

Early pregnancy almost always occurs in the left uterine horn.

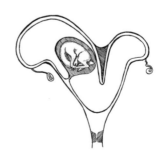

Pregnancy is moderately advanced; limbs and organs of fetus have developed.

With the likelihood of a follicle on either of the ovaries, the female will remain receptive to the male until she is bred. After the birth of a cria, the female's progesterone level drops substantially. This induces her body to produce another egg, usually within 7 to 14 days. She will continue to ripen eggs every 10 days if she remains unbred. When bred several times within several days (often it takes only one breeding), she will release an egg (ovulation). When the egg is fertilized, progesterone is produced by the corpus luteum, the tissue that forms on the ovary where the egg was released. With the presence of the follicle or new fetus in the uterus, progesterone production continues. Gestation in llamas is about 11½ months. It is not unusual for a female to deliver two weeks prematurely or carry her pregnancy to a full 12 months. First-time mothers can carry to 11½ months or be slightly premature. With the approach of birth, the corpus luteum will be signaled by the female's hormones to cease producing progesterone. The cria is born, and the process of follicle production and ovulation begins again.

Copulation

The method of copulation in llamas is unique: No other species of livestock copulates in a lying down position. Male ejaculation is not an instantaneous response but a "dribble," lasting anywhere from 20 to 45 minutes.

The female will kush, with the male mounting her from behind in a sitting position and using his back legs to adjust himself into position.

The male moves forward into position, bending his front legs slightly. He will "massage" the female with his front legs during copulation.

Preparing for Breeding

The dam should be in good general health. She should be properly conditioned and a bit on the slim side. (Overweight females can be difficult for the sire to breed and tend to conceive less quickly.)

Her nutritional needs should be met, paying particular attention to certain minerals (see Chapter 3). All inoculations and wormings should be current. Any infections or chronic problems should be cleared up well in advance, because some medications are harmful to the fetus. There have been some reported incidents of early delivery (10½ months) when the dam is given a typical booster of clostridium C & D 30 days before delivery. But this has not yet officially been directly linked to the booster shot. Moreover, any lameness would make it difficult to withstand breeding positions and the added weight of the male.

The Maiden Female

Just as young males will demonstrate prebreeding behavior, so will young females. A "maiden" female is one who has reached breeding age but has not yet been bred. Large females are not necessarily ready to breed. They have to be mentally and hormonally prepared to accept the male. Some maidens are frightened, reluctant, or confused.

A common fence line or an adjoining stall will gradually introduce her to the male on her own terms. If the male will accept other studs around, then a group breeding will lessen the maiden's fears. She will eventually submit as she sees what other females do.

A young female shows submissiveness to the male by hunching over with the tail up over her back.

Some maidens will actually drop to the ground at the sight of a male, but for the most part, maidens need education and coaxing without undue stress. Using an older, wiser male for the first breeding will sometimes help. These "veterans" seem to know and understand the maiden.

If you are bringing a maiden to an outside breeding facility, have her ovaries palpated by a veterinarian to determine whether she is ready for breeding. If she is not, it is better to know now and postpone her rendezvous.

Most female llamas are capable of reconceiving within 10 days after the birth of a cria. Always examine the female prior to rebreeding to make sure there is no discharge or infection. Infections will hinder conception and may be passed to the stud as well. Crias, however, are unaffected by such infections as they develop after birth has occurred.

Females with Older Crias

It is fairly common for females that have babies from 2 to 5 months old at their sides to experience difficulty in conceiving. This is not a major concern. What's happening is that the female's energy is depleted by milk production, thus hindering the production of healthy follicles. Wean the baby, increase the mother's feed, and it is likely she will conceive again quite readily. (Although the cria will have been weaned early, if it remains with the herd its social development won't be impaired.)

Crias are not normally weaned before the age of 4 months. These babies need at least this much time to get proper nutrients, such as protein, from the dam's milk. They also need to bond with the herd to learn to "be a llama." The most common practice is weaning at 6 months of age. I wean all my crias at 4 months. I find they are mature enough to leave the dams and ready to start training. Refer to Chapter 8 for more information on weaning.

New Moms

The first-time mom can be very nervous. She may show great concern about the welfare of her cria. Attempting to rebreed the first-time mom can be a traumatic experience for her, especially if she is removed from her cria for the breeding. Her anxiety level may get so high that her reproductive system actually shuts down. Keep the cria with her at all times, even throughout copulation if necessary, but use care to watch that the cria is not injured or in the way during breeding.

Nonsubmitting Females

Occasionally a female will refuse to submit to a male, even though you are sure she is open. In some cases, the presence of a retained corpus luteum will make the female think she is pregnant.

Some females, particularly older females introduced to younger males for breeding, simply do not like the "younger generation" and will take longer to allow a young male to breed them or refuse entirely. Some females are just plain teases and like to play hard-to-get. Most such problems can be worked out with the proper introductions, time, and patience.

Oversubmitting Females

Females who are not cycling, maiden females who don't know how to say no, and bred females that go down at the slightest glance from the male are overly submissive. All these different personalities and foibles can be confusing to the new breeder, but experience will help. Get to know your females and their breeding habits. When you purchase a new female, be sure to acquire all the information you can about her breeding habits.

Young or Infertile Females

In some cases, a female may not be ready psychologically. All the right parts are working, but her mind just isn't ready. Give her time; she simply has her own schedule.

Rarely, a female llama may be born without ovaries, causing infertility. This is a genetic defect. The sire or dam who passed on this trait should be culled from breeding. You can determine which parent passed on the trait by observing the results of each one's other breedings. But infertility can also be the result of an infection, nutritional deficiencies, and previous problem births.

Location

Females are sometimes fussy about where they will breed. They dislike having to lie down in wet, muddy, or bumpy terrain. If you are hand breeding (see Chapter 9), the breeder can choose the spot. In wet, damp, or cool weather, a bed of hay or straw will entice the female to stay kushed for a longer period, simply because she will be more comfortable. Try to choose a level area, without slopes. If there is no level ground, face the female downhill to help the male maintain his position.

Specific Veterinary Concerns in the Female

Make it routine to inspect the female carefully after breeding. Note any discharges. Discharges may be semen from the male (milky-white substance) or blood. Bleeding is common with maiden females during the first several breedings, but if it persists, contact your veterinarian. Slight or minimal bleeding for several minutes is common with maidens, but a male will stop breeding if the female has a persistent hymen (a strand of tissue in the vulva). However, this problem is rare. A vaginal inspection by your vet will determine the presence of a hymen, and the vet will know how to treat it.

Pregnancy Determination

It is important in a well-managed breeding operation to determine quickly and positively that females are bred. It is difficult to visibly determine pregnancy in a llama until at least 6 to 8 months gestation and even then it can still be tricky. Llamas carry very well, and usually only older females over the age of 10 years of age will display visible signs of pregnancy earlier, within 4 months. Even the smallest farm can have its yearly profits reduced with the absence of a cria 11½ months after breeding. There are several pregnancy tests available; choose the one you like best. A simple blood progesterone level can be drawn by a vet or experienced breeder and analyzed by a lab. Or you can have your vet perform an internal or external ultrasound.

Behavioral Observation

Most breeders can tell if a female is pregnant just by observing her behavior. It's not a foolproof method, but a breeder learns to recognize the signs. Each female will react differently to pregnancy. Some aggressively spit at the male and actually pick a fight with him; others are docile and continue to allow the male to breed them. Behavioral observation gives only a first indication.

Progesterone Testing

Blood testing for a progesterone assay (test) measures the amount of progesterone present in the female's blood. Typically, any level over 2 nanograms per million of progesterone indicates a pregnant female.

This blood test is done at least 21 days after separation from the male, but preferably at 28 days. If this test is done prior to 21 days, it will merely indicate whether the female has ovulated. At 45 days after separation, the blood test will be able to tell you what you want to know. Female llamas have a significant rate of fetal or embryonic resorption within the first 60 days. (*Resorption* means that the fetal tissues have broken down in the uterus and have been carried away.) Therefore, you may want to perform additional tests at 60 and 90 days.

Progesterone testing is not without its problems. Some females retain the corpus luteum, which occurs when a follicle or fetus dies and the corpus luteum continues working. The female will behave as if she is pregnant — rejecting the male and showing positive progesterone levels — without actually being pregnant.

A female who has difficult births may develop a low-grade uterine infection and display a low progesterone level. This will prevent her from developing a new egg and cause her to be unreceptive to the male.

Progesterone testing is accurate and reliable when used with certain females. Progesterone levels drop in the last stages of pregnancy. If you believe a llama is past her due date, try palpation or, if necessary, ultrasound.

Drawing blood is best left to your vet, but she can teach you the proper technique. Your veterinarian will know where to send the blood samples for testing. Some of the more widely used labs are listed in Appendix B at the back of this book.

Ultrasound

In ultrasound, sound waves reflect off bone, fluid, and tissue at varying speeds, creating an image (in cross-section) of the inside of the body. Internal ultrasound to detect pregnancy can be done at 30 days. A probe is inserted in the rectal tract after all fecal matter has been removed. The veterinarian must be familiar with the equipment and the procedure to obtain an accurate reading.

External ultrasound, used later in a pregnancy, involves placing the probe on the outer left side of the abdomen. While most accurate 50 days after breeding, some veterinarians can confirm pregnancies as early as 28 days after breeding. External ultrasound is less stressful than internal ultrasound, but heavily wooled llamas and alpacas may have to be clipped or shorn in the area where the probe will make contact.

Both these tests require the proper equipment used by a skilled, experienced veterinarian.

Rectal Palpation

Rectal palpation should be tried only on a cooperative female, or on a female who is properly and securely restrained. A restraining chute is recommended. However, some docile females will allow you to palpate while just being securely tied by the head with a lead. By this method a pregnancy determination can be made at 40 to 45 days postbreeding, and occasionally as early as 30 days. Your veterinarian will insert her hand into the female's rectal tract to feel for the fetus and the attendant fluid in the left uterine horn. (The left horn is typically the site of implantation of the egg.) Not all veterinarians have a small enough hand to enter the rectal tract. Some llamas are not anatomically large enough to permit rectal palpation. Your vet can advise you on this.

Mother Nature Knows Best

As I was feeding our girls on a chilly December day, one of them ran out of the shelter, almost knocking me down. But this female kept running in and out of the shelter. I left the feed pans there and went on to the other llamas in the other fields. Upon my return, this one girl was still acting strangely, even though all the others had gobbled up their grain and were behaving as usual. I went inside the shelter and found a newborn cria sitting atop the manure pile, shivering! I went over to check out the cria; it seemed fine except for the shivering. I called to John: "We have a baby!" John was mystified; we had no crias due until May and June.

The female who was acting erratically was visiting, here for breeding, so we knew this wasn't hers. As I was picking up the cria in a blanket to move it to the main barn, its mother passed the afterbirth. This was a new mom who, according to ultrasound a month earlier, was not due until May 15. We dried the baby with a hair dryer and weighed her: 18 pounds. She seemed to be fine, but premature — about 6 weeks premature, not 6 months!

When we had purchased this new mom, in June, the breeder assured us that she was not bred. We specifically purchased this female because she was open — we wanted to breed her to our junior sire, MIV Cabot. Several days after she arrived and had settled in, we put her with our male. Our normally kissy-face male, MIV Cabot, knocked her down, and drove her into the electric fence. It took two of us to pull Cabot away from her. She was moaning and shaking. We removed her and put her in with the other girls in the main pasture. That evening she did not eat and the next morning she was moaning and breathing heavily. We had not had any problem births, but it seemed to me that she was in labor and having trouble. She did eat some grain that night; that made us feel better. I did touch base with our vet, and we agreed to watch her until morning. If nothing changed, he would pay us a visit. The next morning she was fine, and I wrote it off as a case of being overly stressed.

I tried several times to breed her to Cabot, but he would not even accept her over the fence line. He would scream, pin his ears back, and spit. We tried tying him to a post while we introduced her. This girl seemed receptive and even laid down next to him. No luck. He was violent. Our senior sire, Montpelier, was away for the summer breeding at another farm. We decided that when he came back in the fall, we'd try and breed her to him. In the meantime, we pulled blood for a progesterone test, which came back positive. Since some girls will show positive even when they are not bred, we decided that she would have an internal ultrasound.

The outcome? This young lady was 3½ to 4 months pregnant. Shocked, I finally realized that Cabot was telling us that she was bred: He knew all along. We prepared for a spring birth, until I found the little surprise a month later in the dung pile.

I now rely more on the male's reaction to a bred or nonbred female. Since this incident, I have experimented with our senior sire, Montpelier, by bringing to him (outside his paddock) a female bred by Cabot. The same thing happened. Although he didn't attack her, he showed a distinct dislike and spit at her.

The internal ultrasound confirmed the blood test, but I would certainly say it was a little off. Although I will continue to pull blood progesterone tests, I now also pay serious attention to Mother Nature's signs.

Problem Pregnancies and Infertility

Breeders love the female that seems simply to look at the male and become pregnant. Females that "take" on the first breeding and carry a full-term pregnancy are wonderful assets to a herd. Females delivering normally can be bred back within 10 days. Females that have had a *dystocia* — that is, a difficult birth, or prolonged and painful labor — should be allowed at least a month for uterine repair. This condition should repair itself naturally unless there is an infection. Your veterinarian can administer a saline solution for a vaginal wash, and may also insert antibiotics directly into the uterus.

Embryonic Death or Absorption

In normal pregnancies, the embryo will attach itself to the uterine wall within 30 to 60 days. Seven to 15 percent of fertilized ova may die within 60 days of gestation in llamas and alpacas, a higher percentage than in many other livestock. No one knows why so many pregnancies fail. Perhaps implantation failure, stress, hormone deficiency, or embryonic defects accounts for this statistic. No one factor is responsible for problem pregnancies.

Hormone Deficiency

Progesterone is a critical hormone. Production of it may slow down or halt when the corpus luteum malfunctions. Insufficient progesterone levels can cause poor fertility, but an immature hormone system is also a possible factor. A progesterone injection may alleviate the problem. While this is not recommended for maiden females or females under 3 years of age because they may not yet be sexually mature, progesterone has been successful in some cases. If you are considering progesterone supplementation, consult your vet.

Embryonic Mortality and Efficient Management Practices

Waiting 11½ months only to discover that a female has not delivered a cria is a disappointment. Frequent progesterone tests, hand breeding (leading a male and a female to the breeding area and observing), and a thorough inspection of open females will cut the amount of time lost to embryonic deaths. Llama breeders must learn to accept the fact that llamas have a relatively high embryonic mortality rate and to adjust their management practices to accommodate this reality. Some breeders solve the problem by *pasture breeding*, that is, pasturing their females with the male selected for breeding.

Implantation Failure

Uterine infection is the most common cause of implantation failure. Soon after fertilization, the cervix closes to prevent infection. Before it closes, infections can be present owing to a previous difficult birth or unsanitary conditions. Alternatively, the male may introduce infection. Your veterinarian can perform a uterine infusion that, in most cases, will cure the infection. It is always a good idea to have the stud checked for possible infection as well.

In most llama pregnancies, implantation of the ovum occurs in the left horn of the uterus. Occasionally a right horn implantation will occur, but it is thought that the right horn may not provide a suitable environment for the fetus to develop. Mother Nature herself will terminate a pregnancy should unusual cellular division occur.

Disease

A few diseases can cause abortion of a maturing fetus. Toxoplasmosis, brucellosis, and chlamydiosis all may induce abortion.

Toxoplasmosis is carried by cats who routinely eat wild animals or uncooked meat. Feeding cats only commercially prepared foods and trying to keep them from eating wild animals helps to eliminate the problem. The organism is carried through the cat's feces. Llamas or other animals may acquire the infection if the cats defecate near the feeding or watering areas. While more research is being done, it is recommended that populations of cats be kept at a minimum around pregnant llamas.

Most llamas are tested for brucellosis. This disease leaves a llama subject to infection, which in turn can cause the fetus to abort. While not diagnosed in North American llamas, brucellosis has been found in camelids in South America. Imported llamas are quarantined and tested for brucellosis before entering the country.

An organism known as chlamydia can also cause abortion in llamas. While rare, chlamydiosis is diagnosed by testing a fresh sample of the placenta at a laboratory. The organism is passed on from other animals, such as sheep. However, because of the llama's natural immune system, chlamydia is not common in camelids.

The Pregnant Female

Llamas have a normal gestation period of 340 (11½ months) to 345 days, but anywhere from 330 (11 months) to 390 (13 months) days is possible. Behavioral changes in the dam will depend on the dam herself. Many owners of first-time breeding females wonder why their normally sweet, gentle female has become a witch since she became pregnant. This is a shock for many new breeders who have bred and raised their first young female. Other females that are normally grouchy will behave better and even be more manageable when pregnant. Each female llama is an individual.

Llamas are generally easy birthers, needing little or no assistance. They almost always give birth between 11 AM and 2 PM. Mother Nature tries to ensure that the cria is born during the warmest part of the day, has enough time to dry off, and is able to run with the herd at dark to avoid predators. This is a handy schedule in an agricultural enterprise — no sleepless nights waiting for babies to be born. We keep meticulous records to predict when the cria is due to arrive, yet we miss most births. Of all the births on my farm, I have only witnessed two.

Preparation

Depending on her condition (underweight or overweight) and the quality of forage, your pregnant female should be receiving at least 10 to 14 percent protein in her first and second trimesters of pregnancy. She should be getting at least 14 percent during the third trimester. (Refer to the chart in Chapter 3.) The amount of feed will depend on its protein content. Increase the female's rations (primarily grain, but also free-choice hay) by at least one-third during the last trimester of her pregnancy. She is quickly using up the nourishment at this stage in her pregnancy, and she will need the extra nutrition to produce sufficient, high-quality milk.

Most breeders feel that females should be vaccinated 30 days before delivery with Clostridium C & D, but there are vets who advise against this practice due to the stress it contributes. Clostridium C & D is a yearly vaccination required for all llamas, but it is important to give the pregnant dam a booster if she has not had her yearly vaccination within the last 3 to 6 months of pregnancy. Clostridium C & D is a combination vaccination for diseases such as tetanus, enterotoxemia, malignant edema, and blackleg, all of which have been found in llamas. This is added protection for the cria, as the antibodies are passed from the dam to the unborn cria. I have routinely boostered all my females 30 days before delivery with no signs of stress. (See the chart in Chapter 5 for more vaccination information.)

The dam will need a clean and dry place for birthing. Sanitary conditions would include a previously prepared isolated stall in a shed, shelter, or barn, well bedded down. Isolated does not mean total isolation, as this will cause undue stress on the delivering dam. Separated from other llamas but in sight's view is best. Depending on the dam's personality, you may wish to remove her from the rest of the herd a day or two before birthing. A female who normally is quite interactive with the herd and now becomes unsociable may prefer to be alone while she gives birth.

Any females who are due who are normally pastured with a male *must* be

Teats on a nonpregnant mother or nonnursing female.

Teats may enlarge to this stage about 2 weeks prior to birthing.

Birthing is near; teats are full (bagging up).

separated from him. Leaving them together during birthing can be dangerous, as the male will instinctively attempt to breed as soon as she becomes open — even during delivery.

Signs of Impending Delivery

Some females show signs that they are preparing to deliver as early as 6 weeks before their time. Their teats may begin to enlarge, and some females will bag up, that is, the bag will enlarge and fill with fluid from the mammary glands. Others don't exhibit any signs until just before delivery.

As the cervix relaxes to prepare for birth, the cervical plug that keeps the uterus closed will work its way out. This is a mass of mucous membrane, approximately 1 inch in diameter. You may find this on the female's hind end or on her tail. The dam can emit this plug up to 2 weeks prior to delivery or immediately before delivery.

Birthing

Delivery will vary from llama to llama. Some females will become quite restless, getting up and lying down, making frequent trips to the dung pile, or losing interest in food. Other females will behave normally. This beginning part of labor, known as *stage one*, can go practically unnoticed. An obvious sign of impending birth during stage one is up to a 4-inch elongation of the vulva. It may also appear thicker and redder than usual about 1–2 hours before delivery.

In matron females, you may begin to notice that they appear less "sway-backed": The baby has started its journey into the birth canal, which causes the back to straighten out as the weight of the baby leaves the uterus.

Stage two is a more noticeable phase. If you are monitoring your female, you may notice some fluid passing from the vulva.

Proper presentation for delivery is nose and front feet first. Delivery is fairly quick once the head is out, but complete delivery can take 45 minutes or more. Some first-time dams take 2 hours from first labor to delivery.

As the baby proceeds through the birth canal, the oxygen source from the umbilical cord will be severed as the baby's hips pass through the dam's pelvic bones. With the baby delivered out to the shoulders, it now must be pushed out fairly quickly so that the cria may expand its chest to breathe in outside oxygen. You can deliver a baby that appears to be stuck by pulling downward on the front legs. Your experience with livestock will determine how much you are comfortable doing in a problem birth situation. A novice should first call the veterinarian or make arrangements with a knowledgeable local livestock breeder who can be "on call" at a moment's notice.

If comfortable, you may help a stuck baby yourself. A smooth, slow, but firm downward pull can be successfully accomplished by a beginner without causing injury to the delivering dam. If you have any trouble delivering the cria at this point, call your veterinarian. In most cases, the baby will probably be far enough out to breathe on its own while you wait for the vet to arrive.

Problems with the Delivering Dam

Problems with the delivering dam are relatively rare, generally in fewer than 5 to 6 percent of cases. Llama births are easier and more trouble-free than births of any other livestock species.

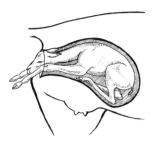

Normal presentation position

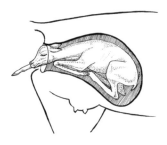

Head with one leg back

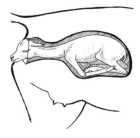

Both legs back, shoulders are stuck

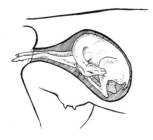

Legs forward, neck and head back

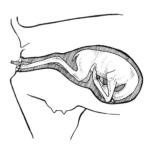

Reverse delivery (breech), hind feet coming out first

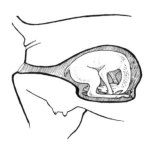

True breech, hind end presentation, no sign of legs

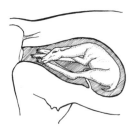

Baby inverted, with all four feet presented at once

Dystocia

Linebreeding is practiced by some breeders. Unfortunately, along with it comes problems. Genetically unsound females will often experience *dystocia*, which means a difficult birth. Some dystocias have causes other than genetics. For example, an overweight female may have a difficult delivery because the birth canal is cramped by excessive fat. A female may experience dystocia only one time or may continue to have difficult births.

Babies that are not positioned correctly in the birth canal will struggle to get through. With an unusually large baby, the positioning may be normal but shoulders can become stuck. A quarter-turn of the baby allows one shoulder through and often corrects the problem.

Contact your veterinarian *immediately* should the baby not be delivered within 2 hours of the start of the second labor stage. In this stage, you will notice some fluid being passed through the vulva. This is from the fetal membranes breaking ("water breaking"). Delivery of the baby should begin in as soon as 15 minutes.

As you become adept at breeding and raising llamas, you will eventually be able to deal with such emergencies yourself. It is helpful to know and recognize abnormal presentations in delivery. Some of the common ones are illustrated on page 153.

Sanitary conditions are important when assisting in delivery, particularly if you need to reposition the baby. With any type of of entry into the uterus, sterile preparation must be made and the dam will need antibiotics to ensure that infection does not set in. Utilization of obstetrical gloves rinsed in Betadine (7 percent iodine solution) will help to ward off infection. The gloves should also be lubicated with obstetrical lubricant, available from your veterinarian.

If a uterine infection should develop, infusion with penicillin suspended in saline solution is helpful. Ask your vet to recommend treatment.

Retained Afterbirth

Do not attempt to pull out the afterbirth if it is partially hanging out. This may cause the uterus to prolapse. *Prolapse* is an inversion of the uterus, a severe and life-threatening condition. If the afterbirth has not passed within 6 hours, contact your veterinarian. Several types of drugs are available that can be administered by your vet to expel the placenta. The preferred drug is actually a combination of oxytocin and prostaglandins. In field births, you may never find the placenta.

The afterbirth is normally expelled 1 to 2 hours after birth.

Twinning

There are only a few documented cases of surviving twin crias. The physical makeup of the dam makes it difficult for her to carry twins to term. The dam may die along with both babies, or only one cria may live. An advantage of early ultrasound is that it is likely to show twins, allowing enough time to make a decision about terminating the pregnancy.

Mastitis

Mastitis (infection of the mammary glands) is uncommon in llamas, but when it does occur, it needs immediate attention. Mastitis may occur when the dam rests in dirty or damp areas. It is a painful condition, particularly if she is nursing. The teat or teats will be enlarged, sore, and hot to the touch. The milk will be off-color. In severe cases, blood will be present in the milk.

Administering antibiotics by infusion through the teat hole is the recommended treatment. This can be done by an experienced breeder or a veterinarian. Check with your vet if you believe your llama has mastitis. Following weaning, use a cattle teat dip to prevent reinfection. A 7 percent iodine dip is more extreme; consult your vet if you think this may be necessary.

If conditions are sanitary and your llamas have a supply of clean, dry bedding, your animals shouldn't have a problem with mastitis. Prevention is the best cure.

EIGHT

CRIAS AND YOUNG LLAMAS

One of the most enjoyable aspect of raising llamas is the cria. *Cria* is a Spanish word meaning "baby." Llama breeders refer to the birth of a cria as *criation*. To witness a birth provides added enthusiasm for this remarkable species.

The birth of a new cria is a herd affair — each member of the herd will personally welcome the new baby.

Concerns in the New Cria

Most baby llamas are delivered without problems, and they will stand and nurse without assistance. Even in a normal delivery you will have a role to play. Sometimes emergencies occur, and human intervention is needed. Any new owner should immediately contact her vet when problems arise.

157

The following are guidelines only. Your use of them will depend on your experience as a livestock breeder. The basic steps should be done quickly, and with as little interference in the bonding process as possible. Your main focus should be to determine if the cria is healthy or to assist the cria until your veterinarian arrives. In general, most dams will allow you to assist or touch the cria within reason. Yet I have a dam that will not even allow her cria to nurse if I am in her stall or holding area. While some females may be outright aggressive, some may not care what you do. As a rule, just let nature take its course.

Put together a "birthing bag." It should include a blanket, towels, OB sleeves, stethoscope, iodine, OB lube, enema supplies, latex gloves, feeding tubes, syringes (including a bulb syringe), and a thermometer. Ready-made birthing bags are available from llama mail-order suppliers.

Gale Birutta

New, healthy crias should be alert. This cria is only 4 hours old, but already his ears are up, his eyes are bright, and he's discovering his new world.

Proper Breathing

First, make sure the baby is breathing normally. Bright pink gums are a sign that the baby is receiving sufficient oxygen. If the gums are white or blue or there is an abnormal breathing pattern, you must make sure the airways are clear. The llama is one of very few animal species in which the mother does not lick the newborn to aid in this process. Wipe any mucus from the mouth and nose with a clean cloth and administer slight suction with a bulb syringe to remove any excess fluid from the nose and mouth.

If you observe signs of breathing difficulty or hear gurgling, turn the baby upside down to allow the fluid to drain from the lungs through the mouth. If this does not help, gently swing the baby in a circle while he is upside down to force the fluid from the lungs. Be sure you grasp the legs tightly and are well away from any obstacles to prevent contact with the cria's head.

If the baby is still not breathing properly, you'll have to try mouth-to-mouth breathing. Put the baby on its side, placing your mouth over the cria's nose and mouth and blow very gently, while watching for inflation of the lungs. If respiratory problems persist or the baby is only breathing with its mouth open, call your veterinarian and take no further action. (Continue mouth-to-mouth until the vet arrives if the cria is not breathing without that assistance.)

Circulation

Provided the cria is breathing on its own and has a clear airway, the bloodstream should be receiving oxygen. Open the cria's mouth and press a finger on its gums, checking for the capillary refill rate. If the gums are pink, the area you are pressing should turn white, then quickly return to pink. If the gums are pale and are slow to color after they are pressed, there may be an airway problem. Repeat mouth-to-mouth breathing, and have someone call your vet.

Life-Saving Experiences

I once had to give a partially delivered cria mouth-to-mouth resuscitation. I was alone on the farm, and when I realized there was a serious problem, I left only long enough to call a neighboring sheep producer for assistance (she was experienced and closer than my vet). The cria was delivered as far as the full neck; the front legs were back and the shoulders were stuck. The cria was suffocating; the airway in the neck was being blocked because the neck was dangling and swinging while the mother thrashed around. I restrained the mother, held the cria's neck straight out, and blew into the mouth and nostrils — keeping this cria alive for 30 minutes until help arrived.

Another time, a llama producer I know performed an extraordinary feat. This remarkable woman revived a clinically dead premature cria on her kitchen table with mouth-to-mouth resuscitation.

While we all somehow learn to perform in life-and-death situations, serious problems are best left to veterinarians or other experienced people.

Umbilical Cord

The umbilical cord will break off during a normal delivery. To prevent infection, you can fill a small paper cup, shot glass, or film canister with 7 percent iodine, available at all pharmacy and grocery stores. Dip the umbilical cord stump in the solution. The iodine will help stop the bleeding and prevent infection. You can stop any excessive bleeding of the cord by applying pressure, tying off the cord ½ inch from the belly, and cutting the cord directly above the tie. The remainder of the cord will fall off within a week.

Drying

Llamas are one of the few species that do not lick their young dry. If the baby is born in warm, dry weather, let nature dry the baby. If the weather is wet or cold, you must towel dry the baby. By rubbing the baby gently with bath towels, you will also aid its circulation. A hair dryer, set on low-warm, can be used in cold weather births, as well.

Temperature

The normal temperature for a cria is between 100° to 102°F. If llama crias are born in hot weather (where temperatures are 80°F and over), they may not thrive as well as those born during cooler months. Hot-weather babies can easily become dehydrated. These dams and babies need a shady area to rest in and drinking water must always be readily available. A hyperthermic (heat-stressed) baby can be immersed in cool water until the body temperature returns to normal. This cria should be checked periodically for assurance that its body temperature remains normal.

Hypothermic (suffering from prolonged exposure to cold, causing a fall in internal body temperature) crias will have low body temperatures and will shiver to create additional body heat through muscle reflexes. Hypothermia causes poor circulation because the blood is kept at the core of the body for warmth.

When a cria is suffering from hypothermia, gradually immerse it in warm water in a large kitchen or laundry sink. The water should be deep enough to immerse the back. Be sure to keep the cria's head up. Its strength may be just about depleted. To increase circulation, gently massage the cria's body all over, especially concentrating on the extremities, while it is in the water. Remove the cria and dry it as soon as the body temperature stabilizes. Keep the cria covered with a towel while you dry it in smaller sections with a hair dryer (not too hot!). A warm enema of water only may also help to regulate the temperature. Or you might also consider using a child's sweater for added warmth. Be sure to put the cria's front legs in the sleeves and roll the sleeves above the cria's knees. (See Chapter 5 for more on hypothermia and hyperthermia.)

> ## Alpacas and Birth Weight
>
> When they are born alpacas are approximately half the size of their llama cousins, and they normally weigh between 10 and 14 pounds at birth. For example, a llama cria that weighs 18 pounds at birth is on the low end of the average scale in terms of birth weight, whereas a 14-pound alpaca baby is on the high end of the average scale.

Weighing

The cria should be weighed as soon as it is dry. This is normally within an hour. The average weight for a cria is 18 to 30 pounds, but a baby can weigh more or less and still be healthy. Weigh a cria on a bathroom scale or with a specially designed scale. You can make one yourself inexpensively from any hanging spring scale with a strap and hammocklike sling to support the cria.

Nursing

The baby will normally be standing within an hour or two of birth, and actively seeking nourishment. It is imperative that the cria nurse within the first 24 hours, preferably within 6 to 8 hours of birth. The newborn cria's immune system is not fully mature. Disease-fighting antibodies are present in the mother's colostrum (first milk). If the baby doesn't get the colostrum, infection becomes a serious threat. Nursing is quick — a few sec-

Gale Birutta

This cria was on her feet within 15 minutes of birth.

An uplifted tail over the back of a cria indicates hunger and will stimulate nursing.

onds on each teat. If you aren't watching, it may be difficult to know for sure whether the baby has nursed.

If you believe the baby has not nursed within the first 8 hours, check for a sucking reflex by sticking your clean finger in its mouth. If the baby sucks, it may just need some guidance. If it does not suck, call your vet. Colostrum may have to be administered via tubing. *Stomach tubing for the cria that is unable to suck should only be done by experienced breeders or veterinarians. A tube passed incorrectly into the trachea can kill a baby.* Don't be overanxious — allow the dam and cria to bond before any intervention, if possible.

Bottle Feeding

You may need to bottle feed if the dam rejects the baby or is unable to produce colostrum (you should suspect this if the cria becomes sick or there is no milk in the mother's teats), if injury prevents nursing, or if the mother is dead. It is essential that the cria receive colostrum within the first 24 hours of life because colostrum provides vital protective antibodies. The first day, the baby needs at least 10 percent of its body weight in colostrum.

If llama colostrum is not available, goat colostrum is a good substitute. Be sure the donor goat has been vaccinated against tetanus and enterotoxemia and is disease-free. The baby should receive colostrum within 12 hours, in four to six separate feedings, with six being the preferred. One pint of colostrum to every 10 pounds of body weight is a good ratio. For instance, a 30-pound baby needs at least 3 pints of colostrum. A 20-pound baby would need 2 pints, and so on. See the chart "Colostrum Requirements by Weight." If nursing naturally, the dam's colostrum supply will normally last only 24 hours and the cria will have had enough colostrum. With bottle feeding, completion of the six separate feedings within the first 12 hours will ensure that the cria has received sufficient colostrum.

There are now colostrum substitutes that are designed for cows. They come in powdered form and are available from feed stores. Check with your veterinarian for her recommendation. For emergencies, have a supply of frozen colostrum. To use it, thaw it slowly (do not microwave) and feed at room temperature. Another alternative is Land O' Lakes Lamb Milk Replacer, which you can get at most farm and feed stores. This is a powdered milk that you can mix fresh at each feeding. Follow the instructions on the package and your vet's advice.

The bottle-fed cria can then be fed goat's milk with a regular baby bottle after you are sure its colostrum needs have been met.

Should the cria not receive enough colostrum in the first 24 hours, contact your veterinarian to arrange for a plasma transfusion. (If you believe the cria is not receiving sufficient colostrum, do not wait the full 24 hours.) Your vet can run a blood test to determine infection antibodies (disease-fighting cells) received from the dam's colostrum. Frozen llama plasma can be secured from Triple J. Farms, 23404 Northeast 8th Street, Redmond, WA 98053, phone number (206) 868-6263.

Colostrum Requirements by Weight

Weight (lbs)	# of Feedings	Total Ounces Needed	Ounces Each Feeding
15	6	24	4
20	6	33	5.5
25	6	39	6.5
30	6	48	8
35	6	57	9.5
40	6	63	10.5

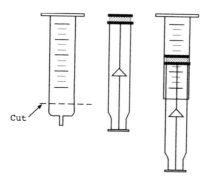

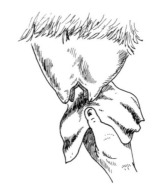

1. A syringe may be cut at the end to make a homemade "pump."

2. Hold a moist, warm cloth over the teats to help loosen the wax plugs.

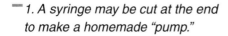

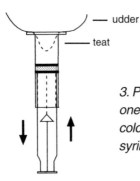

— udder

— teat

3. Press the end of the syringe against the udder with one teat inside the syringe. Pull back on the plunger; colostrum will begin to flow into the syringe. Empty the syringe into a sterile container.

Extracting Milk from the Dam's Teats

If the dam is producing colostrum and she is cooperative, you may "milk" her out to feed the cria that is unable to nurse on its own.

Impaction

Within several hours of birth, the baby should pass the feces that have been retained in its intestines while it was in the womb. These first feces are called the *meconium*. The meconium is sometimes difficult for the cria to pass, because of its hardness and size. If a cria is having this difficulty, it will strain and roll and not be as lively as it should be.

Most often, an enema of warm, mildly soapy water will enable the cria to pass the meconium within several minutes. My farm makes this a standard practice with crias, leaving no question as to whether it has been passed.

Administering an Enema

Mix 4 ounces of warm water with two small drops of a mild soap, such as Ivory liquid. An enema bag or douche bag can be used for this purpose. Lubricate the tip of the applicator with petroleum jelly. The cria may be standing or lying down. Assistance in keeping the cria steady is helpful. Insert the lubricated tip into the anus of the cria not more than 1½ half inches. Steadily and slowly administer the enema.

Diarrhea

Diarrhea in a newborn generally means the dam is producing extra-rich milk. If this is the case, the cria will have a normal appetite and behave normally. This baby is not sick; the problem will usually correct itself when the cria starts on hay or grass in several weeks. If the feces are completely liquefied, you may want to administer a small amount (1 ounce) of Kaopectate orally.

Diarrhea may last from birth to several weeks. This is not unusual. On the other hand, the baby may not be getting enough nourishment and becoming dehydrated. If the cria appears weak or moves slowly, veterinary assistance is required. Fluids will need to be administered along with antibiotics.

Imprinting

Until recently, llama owners were strongly advised against touching young llamas (from birth up to weaning age), particularly males. Breeders and owners were cautioned about overhandling. "Berserk male syndrome" was the main concern. Berserk male syndrome is defined as abnormal and aggressive behavior in adult males who were bottle-fed and overly handled as babies. The bottle-fed male baby often bonds to humans rather than llamas. These males will treat humans as other llamas and, in some cases, become aggressive. Training or castration cannot change this behavior. While I have personally never encountered this syndrome, I have been called by many frantic new llama owners who thought their males were afflicted. New or inexperienced owners commonly confuse this syndrome with a normal male that perhaps has a more aggressive attitude or is lacking training. Since 1987, my farm has been handling young llamas from birth. No behavioral problem has ever arisen. We find that sensibly handled llamas grow up to be more manageable than those who have not received any handling.

Early handling, or *imprinting*, has made its formal debut in the industry. Several experts travel throughout the country showing others how to handle young llamas to bring about manageability. Imprinting is significantly different from general handling. Imprinting involves teaching a cria to overcome its natural fear and to trust humans *within the first hours of life*. This trust is retained by the baby throughout its life.

How to Imprint

Following the normal delivery, allow the cria and mom to interact. Do not disturb the initial bonding process for at least four hours. While the cria is still lying down, being sure not to let it rise, gently massage it around the head, face, ears, mouth, and even inside the mouth. Keep massaging gently even if the cria resists. Each time the baby balks, continue massaging until it relaxes completely.

Continue massaging the cria until you can feel the baby relaxing, but do not stop. Continue for up to another minute around the head, then move on to another part of the body: the back, legs, tail, rib cage, abdomen, and belly.

The entire process can take 45 minutes to an hour, but you don't have to do it all at once. You can repeat the massage several times over the next few days. The results of this massage will stay with the young llama through adulthood, and you will marvel at and appreciate the manageability of this animal.

If the delivery has been difficult, take care of any immediate problems. When the cria is functioning normally, begin the imprinting process. Depending on the problem, you may not imprint your cria for several days, even weeks, if the problem is severe. However, when your cria is well, the imprinting process can begin.

Feeding the Young Llama

Most baby llamas start on hay and pasture as young as 1-2 weeks old. Babies will pick and choose; do not be concerned about limiting quantities of hay and pasture. They will naturally regulate themselves. Avoid feeding them grain until they are at least 3 months old. Babies will not readily eat grain until 4 to 6 months of age; some will not even touch grain until 10 months old. Do not encourage the baby to eat grain before 3 months, as he needs to continue nursing for the high protein that the dam's milk provides. By feeding a baby grain before this time, you will rob him of the necessary high protein. As the dam's milk production decreases, the baby will need other sources of nutrition. Provided the baby has access to quality forage and free-choice minerals, grain should not be necessary other than as a treat.

Once these young animals start on solid foods, they will need an area to feed where they are sure to receive enough forage. A *creep feed* is an area off the main feeding area where they can always find feed. They gain access to this area through a small, low entrance that yearlings and adults are too large to pass through. A creep feed is particularly important if you intend to feed grain.

The Birth of MIV Sheffield

Our MIV Sheffield was born two weeks early. We had installed the electric fence charger on the wall of the barn in Jamaika's (the dam's) stall. During an electrical storm, a bolt struck the fence and found its ground out through the electric fence charger. John and I were inside when we heard the explosion. Running to the barn, we found Jamaika shaking her head to throw off the plastic pieces from the charger. She seemed slightly dazed, but unharmed. The next day she gave birth to a beautiful black male cria. We have never seen such an energetic new cria. To this day we believe he was "supercharged" by the lightning.

Failure to Thrive

While not a common condition, *failure to thrive* relates to a young llama that was born normally but fails to gain weight. These llamas may seem healthy until up to about 6 months, at which time they cease to grow or to maintain their weight. The cause is under study, but some cases involve severe anemia and thyroid conditions. Most affected llamas never reach more than 100 pounds (a healthy full-grown male weighs 275 to 500 pounds) and don't live much past 5 years of age. Research also links failure to thrive to genetic causes. If you have not already bred back the dam, refrain from breeding her if you suspect this condition. While this condition may be genetic (researchers are still unsure) on the sire's side, it is best to cease breeding both the dam and sire of the baby until you research the problem.

Training

When should you start training young llamas? If you imprint, training starts immediately. Some breeders like to wait until the babies are weaned, normally between 4 and 6 months, feeling that the young llama will be less distracted without mom around. Others use mom to help make the halter and lead training less stressful.

Mutual trust and basic training begin simply by allowing the young llama to feel at ease around you. If you have properly imprinted the cria, then the baby is likely to feel comfortable. Going about your chores of cleaning pens or grooming other llamas around him helps the process. Llamas learn by watching other llamas. It is important that they learn not to crowd your "space." Gently push your llama away if he continually crowds you. I like to employ the "three foot rule." A gentle push with a firm "no" will usually do the trick. Don't entice the baby to become pushy by constantly offering him treats. A young llama may be cute when he is searching in your pockets or pulling on your clothes, but this behavior can be dangerous when he is a full-grown breeding male.

Spitting

Llamas will spit at each other during a dispute over food, to establish pecking order within the herd, or to ward off unwanted male attention. It is less common for llamas to spit at people; however, spitting is a defense. A llama generally only spits if he is cornered, intimidated, or being mistreated.

Tips for Halter Training

You may begin halter training when a llama is 1 to 2 months old. If the baby is accustomed to and relaxed in your presence, you will be able to handle it easily. It may balk and rear when it feels the tug of the lead rope, but if another handler leads the mom, you'll be able to train the baby to lead on a loose line by readily following its mother.

Llamas are naturally head shy, and they quickly learn to obey the tug of the lead. They dislike pressure on the lead and will soon learn that when they follow you, the pressure stops. (My babies are halter broken to lead in just two or three 5-minute sessions.) As they approach weaning age, start taking them for walks with llamas other than their moms so they learn that they will not always walk with her.

A friend of mine in southern New England uses a similar technique for training her babies. As a matter of fact, her babies are so well trained that they accompany their moms on obstacle courses in shows at only a few months of age. And they win! It is unusual for females with babies even to be entered in shows, and they almost never compete in obstacle events with their young. This woman is in a class by herself, and I applaud her initiative and training techniques.

Well-trained weanlings make a great impression on visitors (and potential buyers) to your farm. If you plan to show your animals, it is really in your best interest for these llamas to be well trained and taught to stand quietly.

Standing Quietly

It can be a challenge to teach young animals and studs to stand for lengthy periods of time. Find a corner in the barn or along the fence of an outside pasture that provides a solid wall. Begin by tying the llama for 5 minutes at a time. Gradually increase by 5 minutes the time he stands alone until you have built up to 20 to 30 minutes. (Wait until he has mastered a period of time before you look for an increase.) Do not offer grain, hay, or the company of other llamas. The llama needs to learn to quietly stand by himself without a bribe! Some behaviors that might be exhibited by the llama the first time he is tied are humming, wiggling, and fidgeting. Stay nearby but out of sight for the first few minutes until he calms down.

The Young Male

Puberty in any species is a trying time. You may observe males as young as 2 weeks of age mounting mom or other llamas. This is nothing more than play, and not to be taken seriously.

Behavioral changes are the result of hormones as the testicles begin to produce testosterone. Because the penis is not yet attached to the prepuce, a young male mounting females is not a threat because he lacks the capability to extend his penis. (At about age 1, though, a precocious male should be separated from young females to prevent the possibility of early breedings.)

Most males are not sexually mature until they reach at least 2 years of age, although some do mature earlier. Nutrition and genetics play important roles in the early stages of the young male's development. North American llamas seem to come to maturity at an earlier age than South American llamas.

Lithia Llamas

Male llamas are not considered sexually mature until about 2 years of age.

Lithia Llamas

These two crias have just completed their early evening run.

The Young Female

Females may also display instinctive reproductive behavior at a young age. Some females are capable of conception as early as 6 months, but they do not purposefully exhibit breeding behavior until they reach about 1 year of age.

A young female will strut the fence line near young males or geldings. She may push out her chest, walk stiff-legged, and perhaps snort. She may lie down invitingly next to the fence where males are pastured. Perhaps she will curl her tail up over her back and allow other llamas to sniff underneath. (Maiden females after conceiving may also display this behavior.)

Crias as Entertainment

As your herd starts to grow with new crias, their antics will amuse you for hours. Crias will run straight at a fence, and just at the last second, turn and bolt in the other direction, never breaking stride.

Crias will endlessly pester the adults in their pasture. They will jump on members of the herd and tease them unmercifully. Crias particularly like to make a nuisance of themselves with a stud who is temporarily in the pasture to breed. Many breeding studs will tolerate these babies, but be observant.

Lithia Llamas

This cria stayed closed to his dam wherever she went. In my experience, male crias seem to be less independent than female crias.

Llama babies are quite independent — up to a point. From their first day they will wander away from their dams, but they come scurrying back when they find something out of the ordinary — which is just about everything, in the first month of life.

Llama babies imitate their dams in everything they do. When mom makes a trip to the dung pile, baby will, too. If mom decides to take a walk, so will the baby. It is, however, not unusual for dams to leave their babies alone. As long as there are several other llamas in the vicinity of the cria, this is normal and the baby will be fine. Crias are a herd affair. Often, geldings or "aunts" will watch a cria while the dam takes a quiet break or walk.

Gale Birutta

This 2-week-old cria just discovered a stray cat in the pasture. Notice the alert posture: head and ears up, tail up, and chest out.

NINE

BREEDING, GENETICS, AND HERD MANAGEMENT

Proper breeding and herd management will help you meet breeding goals in your program. As a responsible breeder, look objectively at what your farm or ranch is producing and decide which animals to continue breeding and which to cull. Although they may be emotionally difficult to make, decisions to cull animals that are not meeting your program's expectations are a part of good management. The first step, however, is to have a basic understanding of genetics.

Genetics

As in any livestock husbandry operation, you should understand that the offspring inherits genes from both parents to produce the phenotype appearance — how the overall species should look. You will only need to familiarize yourself with the basic genes:

Random Assortment Genes: These are the genes inherited from both parents; genes that generally create the overall, standard or typical appearance of the llama. The phenotype of any species is derived from random assortment genes.

Quantitative Genes: In camelid species, quantitative genes are derived from both the dam and sire, creating a mid-range offspring. These offspring will carry noticeable traits from both parents, such as length, quality, and density of fiber.

Recessive Genes: These are known as "hidden" genes. They can be the livestock breeder's worst nightmare or a pleasant surprise. A mating pair may visibly and physically appear to be a perfect breeding match, but their offspring may carry an undesirable trait, such as a congenital defect. While recessive genes are common in linebreeding, keep in mind that they may, but do not always, cause problems. For instance, two short-wooled llamas may produce a woolly offspring. Researching your llamas' bloodlines will assist you here.

Dominant Genes: These are the expressed genes, or genes that determine the overall characteristics of the species. These genes determine the phenotype.

It is possible to breed an excellent male and female together and get an undesirable offspring — not the norm, but possible. The test of the quality of the male or female is in the offspring it produces. (This is why it is important to view the offspring of any llama you intend to buy for breeding.)

Study mating systems and choose one that is appropriate for your particular management practices and goals. The ideal mating system is an outcross. An *outcross* is a mating between llamas from entirely different gene pools. However, just because an animal is an outcross doesn't necessarily mean that he is better or superior — he should be evaluated in the same way as other llamas. *Outbreeding* is mating from within the same general genetic population, but between unrelated animals. Finally, there are inbreeding and linebreeding. Either of these can create a drastic, rapid genetic change, for better or for worse. *Inbreeding* is mating to a close relative (such as a parent or a sibling). *Linebreeding* is mating to emphasize a desired ancestor in the bloodline. Don't attempt these two systems unless you have thorough knowledge of genetics and have very carefully examined your reasons for choosing either. Inbreeding and linebreeding are responsible for many problems in other species.

Outcrossing

An ideal mating situation is a North American domestic llama crossed with an import. Outcrossing creates increased genetic diversity and adds hybrid vigor (genetic soundness and health) to a breeding program.

While extremely rare, *it is possible* to locate a North American outcross. Most domestic outcrosses have already passed away. Many of these were originally imported from South America by the Catskill Game Farm. Only a handful of these aged Catskill males are still living. Most are found in zoos and are not contributing to the commercial llama gene pool.

Outcrossing holds no obvious disadvantages. However, the breeder must be able to distinguish between a high quality animal and a low quality animal or the outcross may not produce the desired results. Outcrossing is not always better than linebreeding or inbreeding and should be evaluated in the same way you would evaluate any bloodline. Quality depends on proper conformation, health and condition of the animal, and type of fiber or coat desired. Evaluation of an outcross's bloodline starts with the International Lama Registry — obtaining the llama's pedigree. Research involves studying offspring, sire and dam, or siblings.

Outbreeding

Outbreeding is similar to outcrossing. The animals come from within the same genetic population geographically but are unrelated. For instance, another breeder may have an unrelated male that can be bred to your female. Outbreeding is simply another term for breeding within your own geographical region, but with males other than those on your own farm.

Inbreeding

Inbreeding is the quickest way to achieve a specific trait in a llama. It is the mating of very closely related llamas (e.g., father and daughter, mother and son). This may appear to be a good practice, but the disadvantages of inbreeding often outweigh the advantages. When you decrease the genetic diversity, you open the door for problems to manifest in future generations. If inbreeding is chosen, the breeder must be willing to define the amount of risk he or she is willing to take. The norm in any livestock industry is to avoid breeding a pair that share a single grandparent or more. The best decision is to avoid any type of inbreeding.

Linebreeding

When a specific male and female llama are mated together so their offspring remains closely related to the one desired ancestor, it is called linebreeding. The difference between this and inbreeding is that the mated pair of llamas both should be related to the one highly desired ancestor. They are otherwise unrelated to each other. While linebreeding is practiced in many livestock industries, such as goats or cattle, it is risky. The repeated use of one or more desired ancestors may still result in genetic defects, but they will be delayed by several generations.

Genetic Defects

There is no guarantee that a particular breeding program will be free of problems. With a sound selection and mating program, the breeder is actively keeping the chance of defects to a minimum. Abnormalities that do appear in the newborn offspring are congenital defects.

Some defects are caused by the environment; others are genetic. There is a good chance that a defect is genetic if:

- the offspring is from a bloodline that shows a higher concentration of this defect
- other species have been known to show the particular defect to be hereditary
- nutrition and environment appear to be normal

Recessive or "hidden" genes result in defects or traits that appear in an offspring of two sound llamas. Most defects are caused by recessive genes. Defective offspring and llamas producing these offspring should be culled. These same parents may well have produced other phenotypically normal offspring, but the "hidden" gene still remains and will eventually appear again. Some feel that not breeding the same pair to each other will solve the problem, but the recessive gene will continue to be passed on to offspring.

Hidden genes can also produce some pleasant surprises. A llama that is not phenotypically perfect can carry a recessive gene resulting in increased wool, perfect ears, or superior bone structure. Undesirable traits such as blue eyes and tipped ears could be viewed as defects. (Tipped ears can also result from a difficult or premature birth.) Blue-eyed llamas are commonplace in South America and few have shown to be inferior. But in the late 1990s, blue eyes and tipped ears are not considered desirable traits in North America.

Culling your top stud or an expensive female is a tough situation, but the llama industry and your farm will benefit by the decision to stop breeding llamas with undesirable hidden genes.

Breed Groups

The llama industry is in its infancy. Types have been recognized, but not defined breeds. Currently, types of llamas can be briefly described as heavy wool, medium wool, and light wool. These are the show classifications. Packers, pets, and fiber producers are three other classifications. Don't try to breed for all the types. Choose a specific goal for the type of llama you prefer.

Llamas for fleece production should possess dense, high-quality fleece with few or no guard hairs. They should be diverse in color and tolerant of heat. Llamas with shorter, less dense wool are less prone to heat stress. Work or pack llamas should above all be conformationally sound, with good bone structure and an athletic physique.

Pets and companion llamas must have an agreeable temperament, be manageable, and tolerate children. Smaller, gentler animals are ideal; they need not meet the size and conformation criteria of a working llama. It is not known whether heredity determines temperament. Some llamas —

Type Groups of Llamas

Classification	Type	Desirable Characteristics
Show	Heavy wool	Long, dense, and high-quality fiber; heavy bone density; excellent conformation; banana ears; dark in color; manageable personality
	Light to medium wool	High-quality fiber; medium-heavy bone density; excellent conformation; banana ears; preferably dark in color; manageable personality
Packers	Light to medium wool	Sound conformation; medium bone density; tall, with long legs; pleasant and willing personality
Pets/companions	Light/medium or heavy wool; retired packer	Agreeable temperament; tolerant of other animals and children; extremely manageable; smaller in size
Fiber producers	Light/medium wool	Color diversity; above-average quality fiber; heat tolerant; few or no guard hairs; manageable personality

particularly females — seem to pass on their personality traits. But parents with unpleasant temperaments can produce manageable offspring. One exciting aspect of the llama industry is that research is just beginning in so many new areas.

Genetic Goals

In your own particular program, define genetic goals to produce top-quality breeding stock. Primary goals should be excellent conformation, good bone density, overall health, and high fertility rates. Secondary goals should relate to the characteristics you wish to produce: You can breed for size, color, ear shape, disposition, and density and length of fiber.

You'll achieve more rapid progress with a single genetic goal than you will if you set several goals. Genetic changes take up to three generations to

Gale Birutta

"Banana"-shaped ears are highly desirable in the current market.

appear (with the exception of linebreeding or inbreeding). Be patient and consistent. Starting with specific goals will help your farm to be competitive within the industry. Buyers and other breeders will take you seriously if your stock is the result of careful planning.

If you are not a veteran breeder, it may become easy to lose sight of your original goals. The industry is changing rapidly, but don't be tempted to change your goals just to keep up with the current "style." Stick to your plans, but reassess and modify them annually.

Because better-quality llamas are now available, the competitive, knowledgeable market accepts only the best animals for breeding programs. The sooner you cull undesirable animals, the sooner you will have a more uniformly desirable herd. By preserving the best llamas for your breeding program, you head closer to achieving your goals. But when you retire an unsatisfactory sire or dam, do not sell it to an unsuspecting new owner as breeding stock.

Breeding for the Best

Start with a group of unrelated llamas. The llama industry does not have a specific list of desirable traits, so each breeder's ideals establish the norm. With your list of goals in hand, you are ready to begin the selection process. Evaluate your herd for the qualities you want to perpetuate. If necessary, purchase additional animals that may provide what your herd is lacking.

Selective Breeding

It may appear obvious to breed for soundness and conformation first, but some breeders seem to disregard this basic concept. Breeding just for specific traits often results in problems related to health, reproduction, and conformation. Some types of horses, dogs, and goats illustrate what happens when breeders ignore conformation and soundness. In Pygmy goats, for example, breeding for small size has resulted in problems with reproduction. In thoroughbred horses, breeding for speed has resulted in particularly high-strung personalities. Even the young llama industry has experienced this problem to a small degree. However, many fine breeders with excellent foresight realize the dangers of aggressive, selective breeding. Strategies include breeding for superior soundness, along with other desirable traits, to produce llamas much like their foundation ancestors.

The Birth of MIV Cabot

Jamaika was going to give birth at any moment. We pretty much had her due date pinpointed to the day. Every fifteen minutes I checked her; no baby. Feeling safe enough to venture a mile away, I walked to the mailbox. Upon returning, still no sign. The next time I looked out, there was movement in the field. From the kitchen window I couldn't tell if it was the manure pile or a very dark baby. When I went to investigate, it was a new cria!

MIV Cabot was our first llama bred and born on the farm. Our initial reaction was, "What's wrong with him?" He looked like E.T. and had short, straight ears (unlike his parents) and long, curly hair. Although we were quite disappointed (we had studied our bloodlines), we decided that because he was our first, he would always remain with us.

By the following summer, MIV Cabot had blossomed. His wool grew thicker and longer, and his ears took on the distinctive "banana" shape of his sire. He was gorgeous in every way. We decided to take him to the Eastern States Exposition in West Springfield, Massachusetts, to see what the judge thought of him. It was our first show, and we didn't know what to expect.

MIV Cabot clinched second place in the juvenile male division. We turned down four offers of sale that day. So much for the ugly duckling: Never judge your animals until they mature.

The Male

If you have selected or own the best male you can acquire, you have an excellent opportunity to upgrade your females' offspring. The herd sire will have the greatest genetic impact on your herd. A *prepotent* male is one who duplicates himself in the offspring of any female he is bred to. Although breeders may claim that a particular male is always prepotent, this is not so. Some breeders cull the undesired offspring or sell them at weaning age to eliminate any that do not represent their "prepotent" male. Some males are more prepotent than others, but none are 100 percent.

Keep records of which physical type of female is the best cross for your male. A male may cross well with a particular bloodline but not with others. By breeding a single stud to an entire herd of females, you will get an excellent idea of how he is producing. However, you also risk breeding an undesirable trait into your entire herd, and that could set your breeding program back a year or two.

By using several studs, you will be able to choose the one that is producing the best stock by breeding him only to certain females. Having more than one stud does not normally constitute a management problem as it might in other livestock species. While llama studs cannot be housed with each other, they are more manageable than other species. They are easier on fencing and easier to handle. I routinely house studs in adjoining paddocks with no problem. Some breeders house many intact males (breeding and non-breeding) together. In this case, they need to be kept from view and smell of the females.

The true test of a sire is in his offspring. An outstanding male can produce offspring that are not as good as his own phenotype, and an apparently lesser male may produce outstanding offspring that are far superior to their sire. Choosing a stud with no offspring is difficult — you'll have to rely on looks, conformation, and pedigree.

Pedigree alone is not an accurate way to determine genetic superiority. Having a genetically superior and renowned grandparent may be of no value if other ancestors in the line were substandard. Highly promoted and costly animals do not necessarily produce the most desired offspring. Don't assume that the high price of a llama automatically means he will produce quality offspring.

The Female

Few females are being eliminated or culled from the breeding population. Many owners breed any female, regardless of her quality. As the llama market becomes more competitive, these lesser females are sold from a more experienced breeder's herd. Unfortunately, many beginners are anxious to get started establishing their own breeding operation and fail to obtain the proper education and experience in assessing the quality of females. These lesser females are less expensive and therefore more attractive to beginners. By choosing to breed an inferior female that should really be culled, you are deliberately breeding her undesirable traits back into your herd and the industry in general.

Because males are bred to multiple females in a given year, and a female produces only one offspring per year, it takes a longer period of time to measure the genetic capacity of the female. Her importance should not be underrated, as she is still contributing 50 percent to the genetic makeup of her offspring.

By breeding your females to complementary studs, you can create a crossed offspring with superior qualities. Retain the females that are producing these superior babies. Sell the others as pets, not as breeders.

A female may produce a good offspring with one male but not with another. The combination shows up in the offspring. Accurate and thorough records will help you keep track of the crosses that work best with each of your females.

During every birthing season, take a step back and evaluate your herd. What changes and upgrades have been made? Remember, continuing to upgrade the quality of your herd is your first goal. Genetic progress results when you breed intelligently, patiently, and realistically for desirable traits without sacrificing soundness.

Choose the Proper Mate

Breeding for the best entails objectively evaluating your breeding stock and their offspring. Most productive females are used for breeding. But just because a breeder includes a particular female in his program does not necessarily mean that she is of breeding quality. Make a list of your female's good and inferior qualities. Breed her to a male that will improve her poor qualities. For instance:

Female's Inferior Qualities	Male's Superior Qualities
Straight ears	Banana ears
Knock knees	Straight legs and excellent conformation
Short, poor-quality fleece	High-quality, dense fleece
Light boned	Heavy boned

Breeding Stamina

Certain bloodlines seem to produce males that possess superior stamina. Good breeding stamina is important if you are breeding a single male to a large number of females. Strong, aggressive breeding males are highly desirable, especially for hard-to-breed females. A difficult female who refuses to lie down for the male cannot be convinced by a shy, weak suitor.

In females, breeding stamina can be defined as having easy births, the ability to settle quickly, good milk production, and excellent mothering skills. Females are able to produce one offspring per year, with no lapses. Any female who has not had an offspring for more than two years needs a thorough examination by a vet.

Genetic Problems in Alpacas

Alpacas seem to have fewer genetic problems than their llama cousins. In general, alpaca mothers experience fewer dystocias (difficult births) and improper birth presentations. But it could also be that significantly smaller numbers of alpacas in the United States lead us to this conclusion. As the alpaca population increases, it's likely that genetic problems will increase, as well.

Breeding Management and Strategies

There are no wild llamas or alpacas in South America, but the wild cousin of the llama, the guanaco, displays a natural breeding strategy that is utilized by breeders of domestic llamas and alpacas. The dominant male guanaco establishes his territory and guards his herd of females. He drives away all intruders as well as any young males when they reach maturity. Young, mature females are claimed by dominant males in other herds, which eliminates linebreeding. Young males herd together with other young, nondominant males.

Group Pasture Breeding

Pasture breeding has been derived from the wild guanaco's breeding strategies. In domesticity, the breeder decides which male will be the breeding male. Llama breeders thereby eliminate territorial fights among males and assign breeding duties to the male of their choice. With enough separate pasture space, several groups (families, herds) — and thus several studs — can be utilized at the same time.

Group breeding allows the herd's natural social structure to develop. A dominant female will generally keep the male in his place, but will allow him to protect the group and breed other open females.

Advantages

The male is left alone to keep open females bred and will breed any new females entering the herd. This is the best way to reduce stress for breeding females. Considerable space and time are saved with herd pasturing, and the use of more centralized manure piles keeps birthing and grazing areas cleaner.

Disadvantages

Some males lose interest when they are housed with familiar females and may not breed open females as readily. When breeding many females, the male's sperm count will drop, taking several days to recover to normal levels. A male may become sterile due to heat stress, which may not be discovered until later that year. The breeder will only have an estimated date of breeding. Females who have developed uterine infections may go unnoticed and untreated. With regular observation, a breeder should be able to record breeding dates and then predict births with a reasonable amount of accuracy. For overall reproductive herd health, group breeding is the way to go.

Hand Breeding

As the llama industry adopts stricter management practices, owners come to realize that hand breeding provides the most control. It is an advantage to know exactly when a particular male was bred to a specific female.

Hand breeding entails bringing the female to the male's paddock. Both animals are haltered and led. When the male settles into breeding position, you may either stay with the pair or move off to an area where you can monitor progress.

When the initial breeding is complete, remove the female. Reintroduce her 4 days later to check her behavior. If she is very receptive, she has not ovulated and should be bred again. In 50 percent of the cases, your female will conceive on this second breeding. Reintroduce her again in 12 days. Most ovulations last 7 to 14 days.

If the female refuses, reintroduce her again on day 14. You hope she will refuse again; if not, she still isn't cycling. If she is still refusing after three weeks, ask your vet to take blood for progesterone testing or to perform an ultrasound to determine pregnancy.

Wrap the tail of a wooly female with vetwrap, a sock, or a knee-high stocking so that the male will have easier access during breeding.

Advantages

Hand breeding may enable a female to conceive in just one attempt and may help her to maintain that pregnancy, because she is being closely monitored. Hand breeding also lessens the chance of injury to the female. You can inspect for infections prior to breeding and take care of any problems. As the breeder, you choose the site of mating. You can monitor the male while he is breeding to ensure proper positioning and entry. You can schedule births to occur within a certain period, and you lose no time while dealing with an unbred female. Specific due dates free you to attend shows, fairs, and auctions without concern that your animals will need you.

Disadvantages

Human intervention may cause stress to the female. Smaller living quarters, removal from her regular companions, and excessive handling all promote stress.

Farms usually hand breed when outside females are brought in to be bred with a specific male. Females that are held open for more than 60 days to accommodate specific males may be harder to breed back.

Use of Hand and Pasture Breeding

Hand breeding and pasture breeding are most successful when they are used together. Hand breed for the initial introduction to ensure that breeding has taken place. This is particularly useful with a maiden female or a female who is overly protective of her cria. After hand breeding, leave the male and female alone in the pasture for 14 days. If the female did not ovulate after the first breeding, the male will rebreed her.

Use progesterone testing to determine if your female has ovulated. If ovulation has not occurred, retest her every 7 to 10 days while she still lives with the male. When ovulation has occurred, remove her from the male's pasture and return her to her female herd. Retest her again in 21 to 28 days for progesterone levels to determine pregnancy.

Location

Whichever breeding technique you choose, the actual location of the breeding area will be an important consideration. Males are territorial — they are most dominant in their own pasture or paddock. An inexperienced male benefits from staying in his own paddock and having the female brought to his territory.

The female's area sometimes presents a problem. Some studs get easily distracted in the presence of other llamas. A less attentive or young stud may feel confused with many females looking over his shoulder and babies jumping off his back. He may waste time chasing bred females, and occasionally babies may be injured.

With traveling stud services, known studs visit the ranches where the females are housed. For this to be successful, the male must be fairly content. Proper location and conditions will enable him to carry out his duties. Monitor his stress level while he is traveling, keep up his regular feeding regimen, and observe his adjustment to each new place. Of course, some males are aggressive enough that they will mount any female in any location.

Prepare an incoming holding pen or a pasture for visiting females. A traveling buddy will ease stress on the female, but if she has come alone, house her with one of your most docile females until she adjusts to her new location. Housing her in a pen next to the chosen sire will also familiarize the female llama with the stud.

Choosing the Best Breeding Method

Consider each llama's personality, the vigor of the male's libido, your facilities, and the number of females to be bred. Many breeders use a combination of methods to reduce herd stress and keep breeding and birthing schedules on time.

Herd Management

Herd management as it relates to breeding is comprised of culling, castration, blood-typing, organizing the breeding and birthing calendar, and record keeping. Proper herd management will guarantee a smoothly running and profitable operation.

Culling

One of the first steps toward achieving good herd management is to evaluate your herd objectively and to cull those animals that are producing undesirable offspring. (Undesirable offspring include those animals that are not quite what you had bred for, but that otherwise may be excellent animals for someone else's breeding program.) Cull offspring with genetic defects by castrating or spaying.

Ask yourself what your llamas are being used for. Is that male baby potential herd sire material for a top-notch breeding program? Is he potential herd sire material for a breeding program to produce commercial packers or companion animals and pets? Remember your goals. If you are looking specifically for show-quality animals or for breeding stock, then by all means cull the offspring that are of only pet quality. Some potential culls may be outstanding animals that simply do not possess the particular traits you are breeding for in your own program. Evaluate overall appearance, balance, and genetic traits to determine whether to keep or to cull. You may prefer large, athletic llamas; or perhaps you are looking for smaller animals as pets and companions. What are *your* goals for *your* breeding program?

The most important factor in your decision to keep an offspring is whether there is any history of genetic defects in the pedigree or bloodline. Provided there are no defects, the best time to evaluate a llama as a "keeper" is at 6 to 12 months of age. In most cases, legs have straightened, ears are more erect and have taken on their permanent shape, and the condition of the wool is known. (Many heavy-wooled llamas do not display the beginning of their fleece until this age.)

Now that their conformation is more clearly defined, weanlings have more presence at this point and tend to track more correctly, so you'll have a better idea of their balance and how they move.

Male Castration

Castration, or *gelding,* is the process by which an animal's testicles are removed. Castration is performed with a local anesthesia, administered by your veterinarian. The testicles must be descended into the scrotum. An emasculator is used by attaching pincher-like pliers ahead of the testicles onto the neck of the scrotum, cutting the spermatic cord when clamped. Each testicle is done separately. There is little loss of blood and little danger of infection. Castration should take place after 18 months of age. A male that is gelded (castrated) after this time will develop normally. Males gelded under 18 months of age may develop leg and back problems later in life. Gelding is performed to improve the overall quality of offspring. Well-managed farms will reach a higher level of success with their herd and the industry will benefit overall. Over the past few years, there has been much improvement in the overall quality of llamas. More breeders are deciding to geld males that they once would have looked upon as top-quality breeding stock.

Gelding also increases the ease of herd management. For one thing, geldings can remain together or with females and young stock. This reduces your overhead, because you no longer need separate quarters for intact males. Moreover, a male's disposition may improve following castration; thus, he becomes more marketable as a pet or a companion animal. In gelding a lesser-quality male, you may likewise increase his marketability by adding to his utilitarian value as a performance or companion llama. A seasoned, trained packer will bring a substantial price, whereas the value of a potential breeding male may be much lower.

When you make wise decisions about castration, you ensure that you retain only the cream of the crop in your breeding program — and your credibility as a breeder will be enhanced.

Blood-Typing

The International Lama Registry now requires that any males used for outside stud service have their blood types on file. Most breeders consider this a management practice and blood-type their entire herds.

Pedigree accuracy is as important in the llama industry as it is in any other livestock enterprise. Each blood typing possesses individual characteristics, like fingerprints. If the offspring possesses a blood type that is different from either the sire or the dam, the pedigree is probably incorrect. Blood-typing is an accurate tool for determining parentage on either the paternal or the maternal side. Blood-typing also helps to establish parentage by genetic exclusion; that is, blood-typing can also *disprove* parentage.

Blood-typing is useful in several situations. Let's say that a female was bred several times by a particular male and apparently did not conceive. She was then bred to a second male, but the cria did not appear to be consistent with the male to whom she was bred the second time. Blood-typing can determine which male is the sire of the offspring.

Or suppose that a female allows crias other than her own to nurse. Although this situation is unusual, females have been known to "switch" similarly colored crias without the breeder knowing. Blood-typing will determine which cria belongs to which dam.

Although blood-typing cannot correct all errors made in parentage, it is a valuable resource for helping to establish pedigrees.

Information on blood-typing is available from:

Veterinary Genetics Lab
School of Veterinary Medicine
University of California
Davis, CA 95616-8744
(916) 752-7383

Organize Your Calendar

Evaluate and reorganize seasonal breeding and birthing to accommodate your farm's geographical location. For instance, if your farm is in a colder climate — in the Northeast or Northwest, for example — then breeding and birthing should be done in the spring or early fall. In warmer climates, schedule breeding during cooler months to reduce heat stress and possible absorptions and premature births. Study your area's weather conditions in the four months prior to breeding. Conditions must be favorable for breeding, rebreeding, and birthing. Again, the gestational period for llamas is 11½ months, and the ideal breeding program is one cria every 12 months.

Look at the number of producing females in your llama herd and when they are birthing. Are they birthing over a six-month period? If so, reorganize and group your breedings together to have all females birth at about the same time. This will add to your workload for a short time, but it will provide more free time over a longer period. Hold over any females that are not on schedule until the next breeding season, to get them on the same schedule as the rest of the herd. If you purchase maiden females, know when they will be ready to breed so you can fit them into your breeding and birthing schedules.

Group breeding and birthing is far less work and requires less management than breeding at random times throughout the year. Females are more comfortable with other females that are also removed from the herd for birthing. Moreover, weaning is easier with group birthing. All babies can be weaned at the same time, which will reduce or eliminate stress. Crias are less upset by separation from their moms if there are other babies around their own age. Also, crias of the same age will get to play together.

Female Breeding Record

Name_____ ILR#_____ Blood Type Case No._____

DOB_____ Microchip #_____

Date(s) Serviced	Servicing Sire	Comments

Date cria born   Time   Weight

Comments

A breeding record should include general information such as the animal's blood type, date of birth, and ILR number, along with information pertaining to each specific breeding. You may also want to include the dam's weight at breeding, and monthly thereafter throughout the pregnancy.

Record Keeping

Keeping accurate records is crucial to proper herd management and profitability. Every breeding, refusal, or veterinary problem should be recorded. Records for both sire and dam should indicate length of breedings, the level of interest of the male, female unreceptivity, and any breedings that were interrupted. Include other information such as inoculation and worming schedules, and dates of toenail trimming and fighting teeth removal (see Chapter 5). Weigh your llamas periodically and record that information. Document the daily weight of new crias. Record information relating to birthing difficulties, quality and quantity of milk production and colostrum, time of birthing, gestation period, and length of time to pass the placenta. Later you can review these records to better monitor the pregnant female before her next birthing. A calendar or wall chart in the barn is handy for making immediate notes that later can be transferred to the permanent records.

Now that many farms are using computers, there are programs on the market to help you keep good records. These programs can calculate average length of pregnancies, number of breedings required for conception, and the ratio of male to female crias on your farm. There are even some programs specifically designed for llama record keeping written by llama owners, such as Llamatrak and Lamaherd, and various programs from Black Ink Software and Sicon Consulting (see Appendix B for further information). Your hard copy can be individualized sheets, and almost any charts made for other livestock can be utilized as well.

Complete and accurate record keeping will eventually save you time, money, and frustration.

LLAMAS AS A BUSINESS

TEN

STARTING AND BUILDING YOUR BUSINESS

Although raising llamas as pets or companions is rewarding in and of itself, this versatile animal offers numerous opportunities for expanding your hobby into a business. If you decide to develop one or more of the business ventures described in Part Two of this book, you will gain an even deeper appreciation for llamas.

To operate a serious business that will make a profit, you must view it as you would any other business. First, you'll need to acquire the proper identification from federal and state authorities, and obtaining a good accountant can be extremely helpful. Thereafter, advertising and marketing will be ongoing, vital contributors to your farm's financial success.

Legal Requirements

Federal identification number. If you already run another business, you probably have a federal I.D. number. You can utilize the same number by adding a "d/b/a" (doing business as) to your checking accounts and tax forms. To obtain a federal tax identification number, contact the closest Internal Revenue Service Center (the same center where you file your tax return). You will need to obtain form number SS-4 (Application for Employer Identification Number). Even without employees, you need this number to operate as a business showing income.

State sales tax exemption. For tax purposes, contact your state's Department of Revenue to find out what you need to obtain a sales tax exemption number. By filing a sales tax exemption certificate, you may be eligible for tax exempt status when purchasing agricultural products and other supplies connected with your business operation.

Incorporating. You may wish to consider incorporating, especially if you are operating pack trips. By taking this step, you shift personal liability to the corporation. For instance, even though you may have your guests sign a release for injury liability, many releases can be challenged in court. In the case of lawsuit, only your corporation can be legally sued.

Advertising

It is the farm or ranch that advertises consistently through all seasons and during off-peak markets that will reap the benefit of added sales. Don't curtail your advertising during slow months or in a soft market, as this will actually hurt your overall advertising strategy. Keep your business out there all the time to promote it as a strong and steady one.

Advertising Budgets

Budgets for advertising depend on the type and caliber of the llamas you are featuring in your business. Define your market and your advertising needs. Know where to spend your advertising dollars to successfully reach your target markets. If your business involves pet, companion, pack, or guard animals, you can reach your market with several simple ads in local and agricultural newspapers. Businesses involving expensive breeding and show-quality animals have a smaller market to target.

Regardless of your budget, base your advertising on a 12-month schedule. A series of related advertisements over a period of time are more likely to result in steady sales. Consistent advertising also establishes your farm or ranch's name in the industry.

It can be wise to consult an experienced ad designer. Objectively evaluating your own work can be difficult, and if you develop a poor ad, it can hurt your business.

Advertising Effectively

Reach your audience with specific messages rather than general ones. Plan your ads around one concept. This is cost-effective and will bring results.

All ads should include your logo or trademark and your address and telephone number. Typeface, layouts, graphics, and borders should be consistent. Keep the general format of all your ads uniform so that potential customers will recognize your farm or ranch immediately. This will also allow you to concentrate on the message and not have to worry about design. Your main or specific message should be at the top of your ad. It should be short and direct. Bold and italic typefaces are eye-catching.

Photographs and Headlines

If you have a photography background, then you're ahead of the game. (Most amateurs find it difficult to take ad-quality photographs.) These are some guidelines to keep in mind when taking pictures.

Background. Photograph dark-colored animals against a light background and light-colored animals against a darker background. Choose an uncluttered background — you don't want anything to detract from your llamas.

Light. Try to take pictures on overcast days. Early or late in the day is also fine. Always take photos with the sun behind you.

Patience. Set aside a good block of time. Wait for the animal to take that extra step, to turn his head, to put his ears up. Enlist your staff or family to help you get a good picture. One person handles the llama, another gets the llama's attention, and someone else is the photographer. If a stud is the model, someone can walk a female behind the photographer. Your stud will show you his best pose.

Gale Birutta

Light-colored llamas are shown to best advantage in front of a dark background.

The top line, or headline, is also important in drawing attention to your ad. A *great* headline keeps people reading. Some words that capture attention are *success, hurry, remarkable, amazing,* and *bargain.* Look through a livestock magazine to discover what piques *your* interest.

There are many llama videos on the market, but a video produced at your farm showing your stock or packing operation can be impressive. The rule for a "broadcast-quality" (television) video is roughly $1,000 per finished minute. A video production company will normally charge between $400 and $600 for a finished tape. This includes writing the script, final editing, and incorporating sound or music. The fee covers roughly 4 hours of taping, equipment, manpower, editing, and usually several copies of the finished product. Four hours of taping edit into a 7- to 10-minute promotional tape. Don't be tempted to make a longer tape; the average person's attention span is only about 10 minutes.

Endorsements and Testimonials

An endorsement is a powerful sales tool when it is used effectively. For selling either animals or services, a testimonial from a happy customer can bring you lots of business. For example, a breeder who is ecstatic with the offspring from your stud can offer an excellent testimonial for your targeted market. If you use endorsements or testimonials in your ads, be sure to get authorization first and always quote the testimonial exactly.

Diversification Is the Key!

Llamas are in demand as friendly, manageable pets, as companions, and for 4-H projects. They're great for pack trips, which can be highly profitable. Llamas benefit from handling and socialization with different people; packing keeps them (and us!) physically fit. And as the craft market grows, llama fiber is becoming more popular. People who have for years raised sheep for their wool are now adding llamas to their enterprise. These are only some of the options you have as a llama owner. The extent of your diversification can be as broad as your imagination allows.

Farm Tours

Farm tours can be excellent moneymakers. Schedule them for specific days and times. Tours should be booked well in advance. Group tours will show off your farm to many people at once. Perhaps some who visit will want to have llamas of their own one day. Charge a reasonable fee for your time and their tour.

Set up a room in your barn as a display of items for sale: shorn, unprocessed fleeces; spun yarn; organic fertilizer; and other llama-related offerings. Make sure you have prices on everything. Visitors will want to buy the clothes off your back as souvenirs. Farm T-shirts are an excellent idea. You might have something available as a free gift — a baggie of raw wool tied nicely with a ribbon with your business card attached will do. Don't leave these where children can reach them, or you'll be wiped out immediately; you want your visitors to know you are *giving* them a free gift.

As for the tours, they do not have to be formal. You've had visitors before; just be yourself. You won't even have to do much talking other than to respond to questions. Be polite and answer all questions thoroughly, even if you have already answered the same questions 10 times. A sincere and helpful attitude will be remembered, and it adds to the promotion of your llamas and farm.

Make sure there is some time for your guests to have a hands-on experience with a llama — that may very well bring a visitor back as a buyer or a customer for your commercial packing outfit. For interaction, I bring a fresh bale of hay into the female-and-cria paddock and invite the group in. This is quite a thrill for folks who perhaps have never seen a large farm animal except over a fence. It also works wonders with someone who has a fear of large animals. Llamas are ideal for this purpose. Within minutes, people will be asking if you'll take their picture with the llamas!

Contact the folks at your local chamber of commerce for help in promoting tours; they will be most accommodating. Also contact your state's agricultural department or travel/tourist information division. They, too, may have ideas.

Farm Vacations

A farm vacation offers lodging to those specifically interested in experiencing farm life. People will pay well for a real farm experience. Depending on your state's rules and regulations, you may need a license or permit to offer a farm vacation. In Vermont, these regulations are not nearly as strict as those for inns and B&Bs.

After turning away thousands of dollars in lodging requests from hikers taking our pack trips, I decided to do some renovations and offer lodging. It seemed a shame to send everyone who booked a pack trip elsewhere for accommodations. It's been great fun all around: Our guests love the llamas, and we love showing them off.

Of course, welcoming the public into your home means giving up some privacy, but most people will respect your wishes because they have come to your farm and home to experience life with real animals.

Your customers will be looking for a hands-on experience. Guests actually enjoy helping with chores, even with the manure! They love getting "down and dirty" and want to have the real experience.

When pack-trip hikers stay with us, I draw the line at guests "helping" to prepare food for the trips. I found that this slowed me down considerably. I do allow them to help catch, halter, and pack up the llamas for the trip.

We have a small apple orchard with 100-year-old antique varieties. Our guests can pack up a llama and head to the orchard to pick for themselves. Llamas wait patiently while they're loaded up. Further your guests' enjoyment by making an apple pie from their pickings!

Interaction with farm animals, especially llamas, is a delightful experience for all who come. With the llama's calm nature, children and people who have not experienced large farm animals are less intimidated. When they are close to llamas, people ask if they may sink their fingers into the fleece (they may). Baby llamas are especially entertaining as they race around the field.

Fairs

Exhibiting at local and county fairs is not to be confused with formal exhibition at llama shows. Displaying at llama competitions and creating a general informational display at a local fair are two entirely different undertakings. Displays and exhibitions at local and county fairs should be steered more toward education-only displays. Most folks at these functions are visitors and few really have any actual livestock background. Local fairs are always looking for something to bring in fair-goers, so don't be shy in asking for compensation. Anywhere from $100 to $200 per day is reasonable. These displays can be very simple. A live llama on display is your best bet. A simple table with a clean covering such as a tablecloth or sheet with educational literature, fleece, and yarn and articles made from it works well as a display.

Displays at llama competitions should be as professional and thorough as possible. You normally need to buy your space. You are not paid. See more on farm displays in Chapter 11.

Everyone knows what attention-getters llamas are. I realized this early in my llama career. I told the local fair that I could no longer spend the weekend away from my paying pack trips to exhibit at the fair. Sounds selfish? Think about it. I was donating my time at this fair for two days and turning down paying customers for pack trips because I was unavailable. Although I do want to promote my animals whenever possible, I also have a business to run and a living to make. Well, the fair was inundated with "Why aren't the llamas at the fair this year?" As soon as the fair was over for that year, officials contacted me and we negotiated a fee for the next year's fair. The fair makes a profit and besides, everyone else was paid — puppeteers, dog trainers, trick horses. Why not llamas?

That next year, because I was being compensated, I went all out with an educational display. My llamas were well groomed, I set up our show backdrop, and brought a VCR and ran "The Land of the Llamas." This video is a volume in the series "The Best of Nature," a professionally produced series available through Wolfgang Bayer Productions, PO Box 915, Jackson Hole, WY 83001, phone (307) 733-6590. Wolfgang Bayer is a wildlife cinematographer of the highest caliber. My spinner set up an extensive display and spun llama fiber throughout the weekend. Professionally printed brochures from the International Llama Association (ILA) and the Greater Appalachian Llama Association (GALA) were available for the taking. As we were housed in the dairy building, I was surprised by the number of people who tracked us down to see the llamas. Many inquiries came right from within the dairy building from some "old-time" New England farmers. I was pleased with the open-mindedness of most of the "regular" livestock producers. The interest is certainly there, and I have taken bookings for pack trips and sold hundreds of dollars worth of yarn at local and county fairs. Llamas create much interest from other livestock producers looking for diversification.

Store Openings and Parties

Just about every region in the United States has some kind of shopping district. Approach new stores and offer your llamas for "attention-getters." Our farm charges a rate of $100 per hour for two llamas, with a maximum of 3 hours.

Llamas love children and vice versa. I charge the same amount for appearances at birthday parties. However, I do limit that time to 1 hour, as llamas lose interest quickly. Get in touch with party stores in your area. Most of them have contacts with "entertainers" and would be happy to add your name to the list. If there are any marketing firms in your area, let them know of your services.

Television Commercials and Photographers

In 1992, while looking through the local newspaper, I saw an ad for a video production company looking for "actors/actresses" for commercials. I called this company and booked two of my llamas for a 30-second Nissan commercial. The dealer was thrilled with the results and continues to make sales as a result of this ad.

Call local photographers. Make a deal with them to take pictures of children with llamas. This is already done extensively with dogs and ponies, so why not with llamas? Set up a booth at a fair and kids will be lining up to have their pictures taken.

If you are interested in spinning, knitting, weaving, or felting, you may be able to sell the articles you make. Handmade objects also make the most appreciated gifts. Once people become aware of your skill, you may get orders for your handiwork.

If you think about it, llamas can fit anywhere. All you need to do is come up with an idea and pursue it. Many llamas love crowds and behave very well. They are not spooked by cars, running children, or loud noises.

4-H

The 4-H Youth Program is operated by your county extension system. It creates supportive environments for culturally diverse youth in order to help them reach their fullest potential. Traditional livestock producers are becoming more open-minded about llamas in 4-H. The docile nature and manageability of llamas make them usable for children who have not previously been involved with large farm animals. Your farm and animals will benefit, and the kids learn leadership skills and valuable lessons about responsibility. Everyone has a tremendous amount of fun!

Getting a Club Started

You can contact the International Llama Association or your county extension agent to discuss the possibility of including llamas in the local 4-H program. The extension agent will want to know what a child will learn from working with a llama. Extension agents are looking for someone who can promote teamwork, teach leadership, responsibility, decision making, and other life skills. 4-H leaders must be focused on youth development, and not looking for an easy way to promote their own farm or animals. Invite the extension agent to your farm and explain the benefits of llamas and their easygoing nature.

If you have children, it may be easy for you to get a club started. But even if you don't, the Extension Service has newsletters and bulletins to post your project. Many extension agents will be enthusiastic about your willingness to start a llama program. In my area, training and teaching materials are available from the extension agent, and leaders of other livestock groups are an excellent source of support and ideas. The ILA (International Llama Association) has complete packages for starting and running 4-H llama groups. Llamas, kids, and 4-H are a great combination.

When you get the go-ahead, find another adult to assist you in the project. You will probably need help and it's important to have two key leaders in the project. Try to find someone who has experience with 4-H.

4-H Projects

Depending on your schedule, try to allow a minimum of one day per week for the group, but preferably two or three days. Several projects within your 4-H group can be formed: wool, packing, and driving. It is imperative that the entire group learn the basics of catching, haltering, and desensitizing (the process of touching your llama in areas where he is normally uncomfortable until he allows you to do so freely). After kids know these basics, they can choose a project to work on.

Kids can start in the program by learning basic llama management and chores. Establish a graduated performance, with the newest kids first assigned to general barn chores such as cleaning pens, scrubbing water buckets and feed pans, filling grain bins, throwing down and stacking hay, and feeding. After learning the responsibility of chores, they can "graduate" by being assigned to their "own" llama. By the second year, you will have a base of kids who will return year after year. First- and second-year members can be teamed up with experienced club members to teach the kids who have graduated from barn chores to the assignment of an animal. The experienced child will act as the new child's leader and teacher. The third-year project member will be experienced and responsible enough to have sole charge of a llama. As a project leader, you will oversee all activities and monitor each partnership.

Knowledge of fleece and wool should be a requirement for all kids. They can be taught how to care for and harvest a valuable fiber. Assign wool projects. Teach kids how to spin on drop spindles and spinning wheels. The yarn from these beginning attempts at spinning may be used in later projects such as weaving placemats, saddle blankets, or throw blankets. The project members can then use the finished wool projects as money-raisers for their club.

Felting potholders is a favorite among 4-H'ers. After the kids master the basics, the group can graduate to more difficult items such as hats and mittens. Show the youngsters how to add color to their wool projects either with natural herbal and vegetable dyes or with food coloring. Kids can work on these projects and then enter them in the local and county fairs. These kids will receive recognition for a job well done, and their handiwork is sure to sell out.

Never, never sell a child's "project" llama out from under him. This would be devastating. You must make it clear to the youngsters from the beginning that at the end of the 4-H project, that llama is available for sale. The kids must be made to understand this. If a buyer should come to you midway through the season, he must wait for that llama until the 4-H project season is complete.

Setting Up Sessions

All 4-H project members are required to keep a project book. The 4-H'er can custom tailor a book to a specific project, such as livestock, vegetables, or woodworking. Supplement it with hand-out material and worksheets. In the first session, distribute and review the project book and give a general introduction to llamas. Any one of the many fine videos such as "The Land of the Llamas" or "Why Llamas" are available from suppliers (see Appendix C) and will provide an exciting picture of llamas and generate many questions from the group.

The second session can consist of general barn chores like cleaning pens, scrubbing water buckets and feed pans, filling grain bins, throwing down and stacking hay, and feeding. Teach the kids how to approach llamas and how to work around them.

During the third session, kids can learn to catch, halter, and take the animals for walks.

Gale Birutta

Use natural obstacle courses to demonstrate the agility of the llamas and to aid in teaching the youngsters to train the animals.

Additional sessions can include fleece, proper grooming, and shearing techniques. When they know all the basics, the 4-H'ers can then choose an individual project to work on, such as training, feeding, packing, or halter breaking crias.

Win/Win Situation

Both the kids and the llamas will benefit from a 4-H experience — and your farm will, too. 4-H'ers learn a great deal while having fun. Young llamas are handled and socialized. And the farm benefits from the marketing and promotion it receives when the kids show and handle the llamas at public events. With 4-H, everyone is a winner.

ELEVEN

SHOWING

Exhibiting at fairs, public functions, shows, and auctions can be a rewarding experience if both you and your animals are prepared. A fit, well-groomed, manageable llama will be an excellent representative of your operation, and you will derive a great deal of satisfaction from a good performance in the show ring.

The show ring is the ultimate in exhibiting your llamas. Some shows are low-key, but shows sanctioned by the Alpaca and Llama Show Association(ALSA) are top-notch, especially in the Level III category. No amateurs here, these are breeders who have some of the best animals in the industry, along with years of showing experience. Any type of show requires the same basic preparation — make sure your llama is healthy and clean. Beginners are welcome at all shows, but if you are nervous, stick to the smaller, more low-key shows for practice before moving on. Remember that your llama should be manageable before bringing it to any show. Do not use a show as a training ground for your unruly llama.

Grooming for Success

Grooming is a tool for desensitizing your llama. It promotes healthy skin and general well-being and helps in fleece harvesting. It is also a requirement for public exhibition and shows. Depending on the amount of time you are able to spend with your llamas, you will accomplish grooming in two ways: by daily grooming and by keeping your llama from getting dirty.

Preventive Measures

Several factors affect the grooming process, including weather, type of forage, and bedding. For a start, keep your llamas in clean pastures and paddocks. Your pens and pastures should be free of dung, burrs, and weeds to reduce possible exposure to parasites. Feed hay to your llamas in low feeders

12 to 18 inches off the floor — this will keep coats free of debris. Even though overhead feeders are more convenient, debris and forage filter down, falling on the animal's back, neck, and head. This debris becomes embedded in the "oat zone" (the area of the base of the neck at the withers) and is difficult to remove — not because the fleece is denser in this area, but because fleece tends to mat more at the base of the neck where it curves to meet the backbone. (Grain does not present a problem; it is generally fed on the floor in pans.)

At shows, no one type of bedding seems to be better than others; however, shavings and sawdust wreak havoc on a llama's fleece. Wood chips are somewhat better, but they can become entangled in the wool of a heavily fleeced llama. Sand is great for general cleanliness of the fleece, but manure cleanup is a nightmare. Oat straw is a good choice: Its stems do not cling to the fleece, it does not embed itself, and you can remove it fairly quickly with a slicker brush or by picking it out by hand. I prefer straw at shows.

For stabling overnight or transporting for an event, there are sheets made specifically for llamas. These are an excellent investment and will save you time in grooming at the last minute. You can make these cover sheets easily. Fold in half a single flat bedsheet, place it over the llama's back, and cut a slanted break for the neck. The sheet should be hemmed so that it does not exceed 4 inches below the llama's belly. The overall length should be cut appropriately, not extending beyond the furthest part of the topline including the tail. Sew simple ties in two places on either side. When the sheet is on the llama, tie the ties together under the belly. Any type of fabric is suitable. Think about incorporating your farm colors into these covers — this adds to farm recognition.

Bathing

In some llamas, bathing produces a very luxurious fleece. First, a quick surface grooming is required to remove the large pieces of debris from the outer coat. Then use a Circuiteer (blower) to both loosen and blow out the ground-in dirt closer to the skin.

To bathe your llama, wet the coat thoroughly. (This may be difficult, especially if the

Kevin Kennefick

Before bathing, do a quick surface grooming and then use a blower to remove ground-in dirt.

llama has significant guard hair.) Work the shampoo into the coat, paying close attention to manure-stained areas. Rinse *thoroughly*. Then apply some diluted conditioner and rinse a second time. (The conditioner will help release more debris and leave the fleece more manageable.) Conditioners such as Miracle Groom, Elite E-Z Groom, and Show Sheen are all available from general llama suppliers listed in Appendix B. My preference is Miracle Groom, which can be purchased by the gallon for about $30.

There are many coat-grooming products on the market, and they are sold through horse and llama suppliers. Show Sheen, for example, is a detangler that allows debris to be easily blown out of or combed through the llama's fleece when used daily over several days. Such coat-grooming products add luster and manageability. Even white llamas can be brushed thoroughly and blown, and they will look almost as good as if they were bathed. Bathing llamas still requires brushing the coat after the animal is dry, but you have already loosened a major portion of the debris.

While I do not recommend bathing before a show, if you feel you must, bathe your llama at least three days before departing. A bathed llama will need sufficient time to dry out, even with the help of blowers. Barn and stable areas do not have to be heated as shows are routinely held during summer and early fall months. Any touch-ups should be done the day of the show.

Farm Displays

An organized, eye-catching, well-planned farm or ranch display is another reflection of good farm management. Each individual show is different, but most allow llamas to be included in the displays. First impressions and visual images say who and what you are. You may go through several prototypes before you find one that works for you, but all effective farm and ranch displays have certain qualities in common:

- ◆ They are clean, neat, and professional.
- ◆ They target their market.
- ◆ They show that the owner is serious about her business.

If you want to reflect good farm management, you too must be neat, well-groomed, and dressed appropriately for the visitors who come to your exhibit. Appropriate dress ranges from the simple — jeans and a blouse — to the formal — sports jackets for men and a skirt and blouse for women. Again, each individual show is different, so you will need to find out what to expect beforehand. But neatness is what's most important.

Space Limitations

Many shows and expositions have space limitations. Find out before you plan your exhibit how much display space you will have. Keep in mind that space allocations may change — be prepared to improvise.

Design

Display portability, size, bulk, and durability are all factors to consider. The number of llamas and the amount of feed and supplies also contribute to the overall size of your display — and the space you'll need for transporting all of it. A display that is easily dismantled and light in weight will save space in transporting and offer easy setup. You will also have more time to market and show your llamas. Try to enclose your farm display in a designated area, and remember to use your farm's colors.

Display panels should provide a wall for your actual display. You can buy these display boards or you can make them at home. Use plywood — it's lightweight and can be painted, as can latticework. Wood paneling is nice looking but heavier and more costly. Several types of cork and foam boards are also available, but these are less attractive and should be covered with fabric. Also, they may not hold up as well as wood.

Consider whether you want the panels to stand directly on the floor or be table-mounted. By mounting panels on a table, you will have "counter space" for other display materials and storage space beneath the table. Cover the table with a finished fabric or tablecloth that reaches the floor on all sides — again, try to find material in your farm's colors.

What to Include

Farm displays should show what you and your farm are all about. If you are marketing breeding and show stock to other breeders, then focus your display on your top stud and his prize-winning offspring. If you are marketing pets and companion animals, then show how llamas are used with children. Unless your display is a broader-ranged educational exhibit for the general public, don't try to cover all bases; your display will become cluttered and confusing.

Attractive and attention-getting farm banners and 8 x 10 color photos matted and framed on a background are excellent eye-catchers. Special-interest photos can be blown up to poster size for as little as $6 apiece. A very professional-looking double frame with the llama's registration certificates opposite an 8 x 10 color photo is excellent for marketing and selling animals.

Prepare brochures and llama fact sheets for the public. Printed material can get expensive if you hand it out by the hundreds. Many people will take anything available, and the material ends up in the trash. Display only a few copies — this way you will limit the "takers." Store additional copies under your display for distribution to genuinely interested parties. Have a mailing list sign-up sheet — this gives you names and addresses for follow-ups.

Plants and flowers also dress up a display. Even in winter there is greenery available to make your area festive.

A television and VCR mounted in a conspicuous area will draw crowds to your display and keep them there. Commercially produced informational videos or a farm promotional video will greatly enhance your display.

Pens and Stabling Areas

Most shows and fairs provide pens and stabling areas for your animals. If there is no farm display area, you might put up your display panels behind the pen. Be sure they are well secured — llamas get bored at shows and your display could suffer damage at night if they get restless.

Standard livestock panels are routinely used for housing llamas at shows and fairs.

Although it involves a significant amount of work, sewing a backdrop of curtains makes for a stunning display. Remember to coordinate the curtains with your farm colors. A simple way to get a professional look is to visit your local bath store and pick through the various types of fabric shower curtains. You will need to purchase two of the same shower curtains per display panel so that you have enough to stretch across each — the panels average 9 feet.

Many types of shower curtain rings are available, but use plain ones. To adjust the length, clip the curtain to the top of the stall panel and measure to within 1 inch of the floor or ground. Hem the curtain to the proper length for a finished look. Be sure to sew the two curtains together at the side seams.

While exhibiting, remember to sweep the aisles and keep clutter in check. The most dramatic display loses its effect if there is a mess all around it.

Bedding

There are various types of bedding to use when you are exhibiting your animals. Indoor/outdoor carpeting is attractive, but it's also heavy and cumbersome. Keep in mind, too, that it will keep your llamas clean, but will develop urine odors after the first day. A thorough hosing should remove the odor. Rubber mats are comfortable for the animals and keep them out of manure and urine. They can also be hosed off. Though heavy and expensive, they will last a lifetime.

Shavings are the most absorbent bedding and will minimize urine odor. When you change them regularly, shavings are fresh looking and attractive. They're not appropriate for every animal, though. If you are showing a heavy wool llama, this option is a mistake. Llamas love to roll, which embeds shavings in their wool that are almost impossible to remove. Use shavings sparingly with short-wooled or shorn llamas.

Any type of straw (with the exception of oat straw, because of the seeds) works well because it is the easiest form of loose bedding to remove from fleece. You can use straw directly on concrete, grass, dirt, or clay. Some breeders use rubber mats over straw, but this is not necessary. Just keep your area swept — this will confine the straw to the pen and maintain a neat appearance.

Remove manure as soon as possible, and keep bedding fresh and dry. Scrub and fill water buckets often. Always keep your llamas' pens as clean and as neat as possible.

In the Ring

The llama industry boasts some of the finest judges in the livestock industry. When you understand the judging process, you'll better appreciate the job they do.

Understanding the Judging Process

Judges are expected to evaluate to the best of their ability based on what they observe in the ring on that day. Judges abide by llama show industry standards, established by the Alpaca and Llama Show Association.

Ethics. Ethics play the biggest role in judging. A judge must put aside her personal preference for a particular llama type and objectively evaluate *all* animals presented in the show ring at that time.

Consistency. The judge establishes consistent procedure in the ring. Consistency involves examining each animal in the same way. For example, if the judge is picking up tails on males to check for the proper parts, she should pick up tails on *all* males in that class. If a judge neglects this type of consistency, she will quickly lose her reputation for fairness.

Judging criteria must also be consistent. If the judge's first criterion is conformation, then she will place a llama with superior fiber lower than she would one with superior conformation.

Explanation. Exhibitors want a judge to explain why she placed the animals in the way she did. Explanations must be clear and concise, and they must be presented in a positive way. Constructive criticism is useful when it is designed to help exhibitors and their animals. Even the animal that is placed last should be praised for its positive characteristics. The wise judge ends the class with an overall statement lauding the fine group of animals.

Professionalism. Judges must be well groomed and professional. Unfair or degrading remarks by a judge are inappropriate. Communication with any exhibitor prior to the show is not recommended. While many judges will loosen up a bit after the show, it is still best to keep any conversations on show grounds light. Any business to be conducted should be done after the show.

How to Show

One of the best preparations for showing is to attend a showmanship clinic. Llama publications post the locations and dates of clinics in your area. You may also contact the Alpaca and Llama Show Association (ALSA), phone (303) 823-0659, at PO Box 1189, Lyons, CO 80540, for clinic listings and a copy of the handbook.

Young Animals in the Ring

Showing young llamas can create sales for you on show day and can highlight your breeding program. There are three classes for young llamas. Juvenile consists of llamas 5 through 12 months of age; Yearling comprises llamas 13 to 24 months; Two-Year-Olds consists of llamas 24 through 36 months. All llamas 37 months and over are shown as Adults. These age classes are further divided into male and female categories as well as light, medium, and heavy wool animals.

Young llamas must be able to stand quietly for at least 20 minutes at a time. In addition, your animal must allow the judges to touch and examine him. Young llamas can be easily spooked, especially when they are experiencing the confusions of a show ring for the first time. Take these young llamas from their stablemates gradually and expose them to the show facilities individually. Try to familiarize them with show-type surroundings beforehand. Simulate a show stall at home. Confine a llama to a stall by himself, but in view of other llamas. Be sure he cannot jump out. He needs to learn to remain calm for many hours.

These youngsters must also learn to tolerate grooming in the show stall. An unruly animal refusing to be groomed in public does not promote good public relations.

Judges understand the challenges of showing young llamas. They will try to work with you and your animals to help you present them to their best advantage.

Showing Studs

Showing your herd sire can be a rewarding and challenging experience. It is rewarding when he places; it's a challenge because anything can happen. Studs will try to sniff everything. Early training can eliminate this behavior by not allowing the young male to sniff everything of his own accord. Keep your young male in control while in halter and leading, and only allow him to sniff when you say it's okay. (See Chapter 6 for more on managing and showing your herd sire.)

Preparing Llamas for the Ring

To show your llama to his best advantage, you must start working with him long before you enter the ring. You should be well practiced in squaring, or setting up, your llama. This involves placing the animal so that he stands squarely with his weight distributed evenly on all four feet. (See Chapter 6, p. 130, for tips on squaring up your llama.)

In a large class (over 12 llamas), it is difficult to ask your llama to remain squared up the entire time. Keep him in a relatively square position but allow him to relax somewhat. Save that perfect squaring for when the judge is just finishing up with the animal next to you, or for when the judge takes a final look for class placement.

In the Ring

The first objective in the ring is to show your llama to its best advantage in halter. (By learning to handle and show your llamas well, you'll be proficient when you must show llamas to prospective buyers.) You must possess good knowledge of correct conformation and an awareness of what the judge is looking for as she moves around your animal.

Remember these positions by the phrase *back side, same side; front side, opposite side.* These positions are for the safety of the judge. By keeping these positions, you have more control of your llama as the judge makes her way around the animal. Back side, same side allows you to move quickly between the judge and the llama should the llama spook or kick.

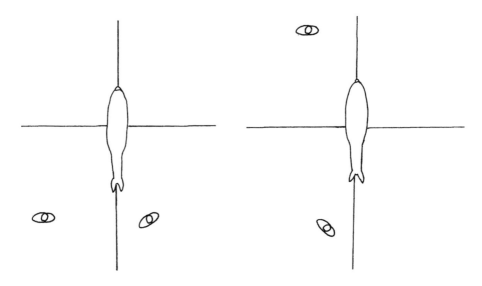

1. This is the "show position." Keeping his toes pointed toward the llama's eyes, the handler faces the llama at a 45-degree angle off the llama's shoulder. The handler stands in front of and off to the side of the head. At this point, the judge is looking at the llama's head, jawline, ears, and eyes. A complete viewing of the llama's front is performed at this stage.

2. The judge will move to the rear of the llama, pausing at the llama's side, staying on the same side. Slowly the handler must move across the front of the llama, to reach and stand in a position at a 45-degree angle on the other side of the llama's head. The judge, by moving from the front to the rear, is evaluating the llama's topline, leg placement, and overall balance.

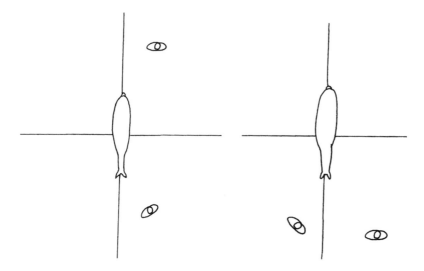

3. (Above) As the judge passes behind the llama, she is looking at the rear of the animal — that is, she is examining for straightness of legs from a rear view. The judge will ask if she may pick up the tails on intact males to check for the presence of testicles. (It makes sense if a male is entered in an intact male class that he be equipped with proper breeding paraphernalia.) In competitive and close placing decisions, judges may place the male with larger testicles higher as indicating a finer example of a breeding male.

4. (Below) As the judge moves to the front, she will ask if she may touch your animal. This will be a hands-on check of the llama's topline. (Topline is the line that runs from the base of the llama's neck to the top of his tail. A llama should have a straight topline and not be rounded or sway-backed.) As the judge continues to move to the front, slowly retake your position at a 45-degree angle as shown in the illustration below.

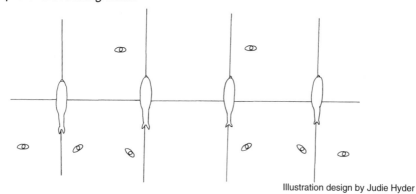

Illustration design by Judie Hyder

WHAT THE JUDGE IS LOOKING FOR

Head and Neck	Alert, uniformly shaped ears; bright, wide, well-set eyes; proper jaw alignment and normal teeth.
Body	Straightness of topline, with a tail set that continues straight off the end of the topline (tail set is the positioning of the tail in relation to its attachment to the llama's body. A too-low tail set might indicate some cross-breeding with an alpaca); proper conditioning and muscling; proper distance between front and rear legs; four teats on females; two uniform, visible testicles on males.
Legs	All legs should demonstrate good bone density and straightness. All toes should face forward and pasterns must be properly angled. Judge will also look for properly trimmed toenails and clean feet.
Fleece	Uniform density, fineness of fiber, and general overall health and luster. Grooming will also be scrutinized.
Balance and Way of Moving	Gait should be fluid, graceful, and without short, choppy steps. The entire balance of the llama should be well-proportioned and symmetrical. Most judges find movement extremely important in placing an animal. Some llamas may appear to be knock-kneed while they are standing but in fact have excellent movement. This shows the judge that they are proper in their conformation. Most of what the judge needs to see is evaluated while the llama is moving.
Attitude or Disposition	A manageable, aiming-to-please attitude.

Tip

Remember: Always stand at a 45-degree angle with your toes facing the llama's eyes. In this position, you are able to watch your animal and the judge with just a quick turn of the head. When the judge is at the front of your animal, you should be in the opposite quarter. *Never, never* get between the judge and her view of your llama.

Halter Classes

Judges evaluate conformation in halter classes. But here are some additional guidelines for showing in halter classes:

1. There will be a ring steward to assist the judge. The steward will tell you in which direction to circle the arena, where and when to line up your animal, and when to pull your llama out of the line. If this is your first show, the ring steward will guide you through the pattern. Judges have different patterns for entering the ring, lining up, and even placing, but these are discussed with the ring steward prior to the show.

2. Enter the arena clockwise at a brisk walk. First impressions will stick in the judge's mind when your llama is moving out freely. Lead your llama on a slack line from his left side, next to his head.

3. Keep a reasonable and safe distance between yourself and the next llama, particularly if it is a stud class. With the growing popularity of llama shows, more and more animals are entering competitions. You may have to deal with a crowded ring. Try to keep 6 to 8 feet between animals.

4. Lead the llama from the left-hand side with the lead in your right hand held at least 8 inches from the halter. Your left hand should hold a figure-eight coil of the remaining lead, at waist-height. Do not pull down on your llama's head. This will throw off his balance and make him hunch his back. Hold your llama's head up and he will deliver a proud carriage.

5. When lining up, remember the 6-to-8-foot rule between animals. Again, this will depend on the size of the class. If you are showing a stud that is antsy, ask the ring steward

beforehand if you can enter the ring last. This will reduce the time needed to keep him still in the arena as others enter.

6. When the judge asks you to leave the lineup, walk briskly toward the judge as directed by the steward. Do not block the judge's view of your llama; the judge wants to evaluate the balancing and straightness of your llama from the frontal movement. Make sure the judge has a clear view of the llama's line of travel! You may be required to stop at the judge or to continue around the arena.

7. Be prepared to have the judge handle your llama. The judge may feel for bone structure or straightness of topline. Quickly and discreetly smooth out the fleece on the llama's back after it is ruffled by a hands-on check. A llama that allows a judge to handle it will have an advantage over the llama that will not. A superior animal may place under an inferior one only because the judge couldn't evaluate it by a hands-on examination.

8. When turning, rotate the llama away from you. Push the llama's head away to circle or turn. This maneuver will cause the llama to turn on his haunches, called a *pivot turn*. The llama should be on the *inside* of the turn.

9. When returning to the line, lead your animal through the line and pivot turn in back of the line. Never turn your animal in the lineup — you may bump into and disturb other llamas that have already been set up.

10. If your llama is acting up and is becoming uncontrollable, swallow your pride and remove him from the show ring. You have nothing to gain by staying, and you don't want your unruly llama to disrupt the others.

Showmanship Classes

Showmanship is not to be confused with actual showing. In showmanship classes the exhibitors themselves are judged and placed, not their llamas. The exhibitor is judged on how well he presents himself and his llama.

When you are in a showmanship class, you will be judged on how neatly your animal is groomed and on your own grooming. Your overall "professional look" and that of your llama are judged here. Sanctioned shows are strict on dress code for the handler. Men wear black slacks and a white shirt; women must wear a black skirt or slacks with a white blouse. This way everyone presents a uniform appearance in the ring, and there is less to distract the judge from evaluating the animals.

A good showman must be poised, alert, and exhibit good sportsmanship. Smile. Keep one eye on your animal and the other on the judge. *Remember, you are showing at all times while in the arena!* Be aware of the other llamas in the ring; be prepared to move if they crowd you. Displaying and practicing good showmanship will help your animals place higher in a halter class.

Obstacle Classes

Most obstacle classes are more relaxed. Sanctioned shows will still have a dress code, but the rules may be more flexible. Obstacles include bridges, ramps, and stairs. The course may include jumps ranging from 16 to 24 inches in height. Water obstacles will contain a minimum depth of 4 inches of water. Pack obstacle classes will include deadfall consisting of posts, logs, and tree limbs. Handlers will be required to back their llamas between hay bales or other objects. Llamas will encounter such other unfamiliar objects as wheelchairs, barking dogs, or bicycles. Obstacle classes will also include loading in trailers or vans.

The best preparation for obstacle and pack classes is to practice at home. Many times a llama that performs well in a show ring obstacle class has never had any real show experience but has practiced at the farm. A particular packer that I used in my commercial string, MIV Jericho, was by far the best field packer I ever had. Thinking he would easily win the obstacle class, I entered him in one. After the second obstacle, he lay down and looked at me as if to say, "Come on, Mom, get real — this is kid stuff!"

ALSA has specific guidelines for obstacle and pack classes and puts out a handbook covering the specifics. Write to ALSA, PO Box 1189, Lyons, CO 80540.

In obstacle classes, judges look for communication between the handler and the llama and how much that llama trusts and responds to its handler. A llama will lose points if he refuses an obstacle.

Success in the Ring

A high placing in the show ring makes you feel terrific. Follow the guidelines above and attend sanctioned ALSA clinics to sharpen your skills. Practice at home, too, and you and your llama will be on your way to consistent wins.

By mastering the art of showing and handling your animal, you will be displaying your llama to his best advantage. But as a bonus, you'll also be displaying your ranch or farm in a professional light.

Impeccable grooming is a top priority, along with properly fitting tack. (See Chapter 13 for more about grooming.) Halters mustn't be too tight around the nose and not so close to the llama's eyes that the llama squints. Tack used in the show ring should be reserved exclusively for showing. Clean it and store it when it is not in use. However, pack equipment is also utilized on the trail, so show officials are lenient regarding the condition of this gear. It is expected, of course, to be clean and in good repair. Within the requirements of ALSA rules, halters and leads should be plain black leather or nylon, with no dangling attachments.

Remember, just because your llama does not place as high as you expect at any given show, this does not mean you have an inferior animal. It just means the judge evaluated what she saw in that ring on that day. Don't be discouraged; your llama will place differently at other times, at other shows, under other judges. Work with your llama and follow the techniques you've learned, and you will ultimately have the competitive edge in the arena.

Ribbons, Awards, and Recognition

No one can deny the thrill of winning awards and ribbons. Such recognition brings a personal sense of pride and satisfaction as well as a heightened reputation for your breeding program and herd.

Ribbons and Awards

The larger the show and the more entries, the larger the winnings. ALSA-sanctioned shows judge llamas from 1st through 6th place. First place may be a blue rosette ribbon with a purse of $75; smaller shows may award a blue 1st-place "bookmark"-type ribbon with no money.

The most honored and prestigious award at any show is Grand Champion. Some advanced and established shows provide a flowing rosette ribbon, usually in purple, in addition to a trophy. Reserve Grand Champion is second only to Grand Champion and may also receive a rosette and trophy.

Cost

As not all llama shows are sanctioned by ALSA, entry fees range from no charge to as much as $100 for high level and advanced shows. Even shows that are sanctioned by ALSA may not charge entry fees when they first start up. This is to establish the show and encourage as many entries as possible. These fees range later from $5 to $25 per llama. Entry fees are per llama, not per class. For instance, three llamas may be entered in a total of nine classes. The cost then would be $15 total — $5 for each llama. If you are competing in ALSA-sanctioned shows, you must pay an annual membership fee and your llama(s) must be registered with ALSA, or additional fees may be requested if you are not a member.

While you won't get rich entering and winning at shows, your farm receives considerable recognition if your llamas do well. My llamas have come away from a successful season with as much as $1,500 for six shows. While a llama may not become more valuable financially, your breeding program will benefit from his success, and you will likely make more sales of young llamas. Some farms raise stud fees considerably when their offspring show well.

Showing does entail work and investment, such as travel expenses, entry fees, and any additional stabling charges. And you'll have to pay for veterinary health certificates that are required for moving llamas from state to state. Different states and different shows require different tests. If you are considering a specific show, ask for a show book, which will spell out state and show test requirements. New Jersey, for example, does not require a health certificate, but you will need to have testing done anyway to re-enter your particular state. Your veterinarian will know what tests are needed for the state to which you are traveling.

Show Levels

Various show levels are established by ALSA:

Level I Shows (75 llamas or fewer entered)
Mandatory Classes
>One complete halter wool division
>Open Obstacles

Level II Shows (75–175 llamas entered)
Mandatory Classes
>Two complete halter wool divisions
>Nonbreeder, Get-of-Sire, Produce of Dam
>Open Obstacles
>Open Public Relations or Open Pack and Young Pack
>Youth Showmanship, two ALSA age divisions
>Youth Obstacles or Youth Public Relations

Level III Shows (more than 175 llamas)
Mandatory Classes
        All three halter wool divisions
        Nonbreeder, Get-of-Sire, Produce of Dam
        Advanced and Novice Obstacles
        Advanced and Novice Public Relations
        Advanced, Novice, and Young Pack Llama
        Performance Champion and Reserve
        Youth Showmanship, two ALSA age divisions
        Youth Obstacles, two ALSA age divisions
        Youth Public Relations or Costume

Youth Classes

At all shows, small or large, sanctioned or nonsanctioned, you will find youth classes. These are usually the entertainment highlight for spectators. In ALSA halter classes, the llama receives the points and not the handler. At llama shows you will see youths and adults showing at halter while in the same show ring. Youths actively competing in ALSA shows must meet the same requirements regarding membership and registration of llamas as adults do.

Presently, classes for youths are showmanship, public relations, and obstacles. New for 1997, ALSA has added youth packing classes. In Level III shows, costume classes (both llama and youth handler) are extremely popular and will draw a crowd. In these classes, the youths receive the points and not the llama. Many youths do not own their own llamas; they show other people's llamas, or receive a new 4-H llama every year. 4-H Clubs, Future Farmers of America, and private youths may show.

The Youth Division is divided as follows:

- ◆ 7 Years and Under — Sub-Junior (not ALSA approved)
- ◆ 8 through 13 Years — Junior
- ◆ 14 through 18 Years — Intermediate

Where to Find Shows

It's easy to find shows to enter. Provided you have joined your state, regional, or international association, you will receive mailers and flyers or publications advertising shows in your area, as well as those held further away.

More Common ALSA-Approved Halter Classes

Wool Class	Sex	Age	Placings
Light-medium wool	Female	Juvenile 5–12 months Yearling 13–24 months 2-year-old 25–36 months Adult 37 months and over	1st–6th each class*
Heavy wool	Female	Juvenile 5–12 months Yearling 13–24 months 2-year-old 25–36 months Adult 37 months and over	1st–6th place each class*
Light-medium wool	Male	Juvenile 5–12 months Yearling 13–24 months 2-year-old 25–36 months Adult 37 months and over	1st–6th each class*
Heavy wool	Male	Juvenile 5–12 months Yearling 13–24 months 2-year-old 25–36 months Adult 37 months and over	1st–6th place each class*

Group Class	Requirements	Placings
Produce of Dam	Two offspring by the same dam (any sex); dam not present	1–6
Get of Sire	Three offspring by the same sire (any sex); sire not present	1–6

**In each of the above classes, for each sex, 1st-place winners from each light-medium wool class and heavy wool class return to the show ring for placing for Light/Medium Wool Reserve and Grand Champion Female, Heavy Wool Reserve and Grand Champion Female, Light/Medium Wool Reserve and Grand Champion Male, and Heavy Wool Reserve and Grand Champion Male, respectively.*

Shows are sponsored by llama groups nationwide and are normally held at local, county, and state fairground facilities for convenience in conjunction with area fair schedules. If you would like to start a show in your area, check with the ALSA or the board members who run your local fair.

TWELVE

PACKING

An advantage of owning and raising llamas is that they have utilitarian value as pack animals. Bred for centuries for this purpose, farm-raised llamas can readily adapt to their ancestral role. (Today, you'll still find llamas used in remote areas of Peru and Bolivia where other means of travel and transportation are not possible.)

Llamas in North America have served their owners well since the 1970s. During that decade, hiking with llamas was popular recreation and the llama population in the United States started to grow. For those who enjoy the wilderness and backcountry, pack llamas make the journey easier. An increasing number of recreational hikers and campers now own or lease one or more llamas to use as their personal packers. In some areas of the country, the U.S. Forest Service now uses llamas in lieu of horses or mules.

With more emphasis on the environment, more individuals and families are choosing to "get back to nature" with outdoor activities as a vacation. Llamas provide companionship, and llama outfitting guides are generally knowledgeable and educated about the environment.

Depending on the duration of the trip, time of year, and if meals are supplied, trips can range in price from $25 per person for a half-day trip to as much as $1,000 per person for a week-long trip. Llama packing is most popular in the Northwest and is become increasingly popular in all areas of the United States. Choosing to enter the llama packing business is enjoyable, self-satisfying, and lucrative. Using llamas for one of their original intended purposes also helps in the promotion of the species.

Commercial Packing

The type of packing you do depends on the number of llamas you have or plan to purchase, how much time you can devote to managing the business and spending on the trail, and your budget.

Lightweight Packing

Lightweight packing can involve day trips or half-day trips. They can be adventurous hikes or easy, leisurely hikes to enjoy the scenery and the company. You can serve picnics or gourmet spreads. Lightweight packing is an excellent way to break into the commercial business: It requires little time and money.

Because they may be involved in shorter trips, lightweight pack llamas won't need as much physical conditioning as heavyweight pack llamas. You must decide how many people you can accommodate. The more people, the more food you will need. Two llamas should be sufficient for four to eight people. Fewer animals means that you will be able to transport them in a truck or van, without the expense of a trailer.

Heavyweight Packing

Heavyweight packing requires more time and preparation. These types of trips generally involve at least one overnight, but some may be 10 days or more. Depending on where you're going, it could take a day just to get to the trailhead. The financial commitment increases as well. You may be accommodating a larger group and will certainly carry additional supplies. You'll need more pack llamas, and they will have to be well trained and seasoned (seasoned llamas are ones that have a year or two of experience in packing).

Other investments include the purchase of tents, sleeping bags, and camp kitchens. And you'll be using additional help as well — it is impossible for one person to handle commercial packing with a large group on a long trip.

Starting Up

Commercial packing can reap excellent financial returns. To get started, you must secure animals. If money is no object, find the best already-trained and seasoned commercial packers. Commercially operating llama outfitters are reluctant to sell a well-trained packer, but you might entice them with a good offer. If you are not in that enviable financial position, take in inexperienced llamas to train as packers, or purchase llamas 2 years and older that would make good packing prospects.

Summer "Pack Camp"

Larger llama breeding farms sometimes have several intact males or geldings that are for sale at a reasonable cost. Approach one of these breeders and offer to take some males off his hands for the season. Intact males make great packers if there are no females around. The breeder may be happy to ship off males or geldings for a "Llama Summer Pack Camp." Offer free board and field pack training, which will both benefit the farm and increase the animal's value. The farm lending the llamas would receive a fit, trained packer at the end of the season. Although you would be starting with untrained packers, you wouldn't be investing much money.

Contact breeders in late winter and line up potential packers — when summer comes you'll be ready to go. (By starting early, you may even have to turn away breeders who like the idea.) You'll have plenty of time to advertise your trips and promote your animals and your farm.

You'll need a written agreement, which you and the llama owners must sign. Be sure to cover who is responsible for upkeep, injury, illness, and even death. The agreement should spell these items out. A simple contract can be drafted similar to a purchase and sales contract, modifying the same for free lease (a lease where the animal is supplied at no charge) use. Don't forget insurance on the visiting llama; your farm insurance may not cover visiting animals. With my farm, I cover all expenses such as veterinarian, feed and grain, supplies, and equipment. My farm is also responsible for any injury to the visiting animal while it is in my care.

Selecting the Correct Pack Llama

Males are traditionally used for packing, although females can and do make excellent packing companions, they are rarely used because of their breeding value. Gelded males are the best choice.

Conformation

Not all llamas make good packers. The ideal pack llama should possess excellent conformation. Look for straight, strong legs, set well under and in proportion to the rest of his body. A narrow-chested llama is likely to be a better pack animal than a broad-chested one simply because less bulk means quicker movements and greater endurance. For the lightweight packer or family packer, almost any conformationally sound llama will do, provided you limit his work to several trips every few weeks with loads of 50 pounds or less.

Wes Holmquist and his business, the Llama Connection, are considered the industry leader in training and packing with llamas. Wes has packed and trained more than 150 llamas. In 1993 he established the Classic 2000 Registry, which registers only llamas that meet certain requirements. Wes recommends 10 structural guidelines for commercial packers:

Dennis Jensen

A llama with a fluid and smooth gait will be able to move out quickly, without short, choppy steps.

1. Llamas should be over 43 inches at the withers.
2. Llamas should have longer legs than depth of body. Deep, wide bodies are not a plus (depth of body should be less than 21 inches).
3. Llamas should have a level topline. It is common but not desirable for a llama to be shorter in the shoulders than in the hips. For packing purposes, a llama shorter in the shoulders will endure undue physical stress while descending. A llama shorter in the shoulders than in the hips is conformationally not correct and should be reserved for pets, companions, guards, or 4-H projects.
4. Llamas should be under 6 inches between the forelegs. It is a disadvantage for pack llamas to be wide between the front legs because it becomes more difficult for them to work their bulk.
5. Llamas should have a narrow-to-medium frame.
6. Llamas need evident chest muscling to tie forelegs and shoulders to their chest.
7. Llamas should not be gelded before 18 months of age. Gelding before that age will affect their hormonal growth, and their bones will grow poorly.

8. Llamas must have a long, free stride with their front legs for endurance on the trail.
9. Llamas must have a good general conformation with strong ankles and pasterns.
10. Llamas should not be over 425 pounds. Extra weight is a disadvantage for pack animals. (See Chapter 3 to learn how to estimate a llama's weight.)

The Ideal Structural Type

Wes Holmquist identifies the five ideal structures for a pack llama:

Muscle Llama: *"Medium-width frame, big bone, big muscling, chest muscling ties into forearms, good ground clearance, medium-length back."*

Full-Sized: *"Big bone but narrow frame and excellent ground clearance, wide and long knee joints, medium-length back."*

Tall: *"Narrow frame, long legs, medium bone, medium-length back, chest muscling evident. Tall llamas must have a shorter back proportionally than medium-sized llamas."*

Medium: *"Narrow frame, medium height, a lot of ground clearance, medium bone, but longer back."*

Compact: *"Shorter wither height, medium to large bone, longer legs than depth of girth. Can be big muscled and wider proportionally between the front legs and still be light, athletic, and have good endurance."*

Attitude

A llama's attitude is almost as important for packing as his physical structure. While a willing packer — a llama with "heart" — is desired, the potential pack llama must be ready to work with you in the areas of catching, haltering, and grooming, too. An enthusiastic llama with lesser athletic ability is a better choice than a llama with superior physical capabilities but a poor attitude. Once you have trained your llamas, they will be working daily, carrying heavy loads, during the long packing season.

There is one other important consideration: Your llama should allow you to pick up his feet. In case of injury on the trail, he must be cooperative and tolerate any examination you might have to make of his feet.

Training

Choose a potential pack llama that leads easily and politely. The average llama walks at a slower pace than most hikers and walkers, so be prepared for the llama to slow you down. Ability to lead on a slack lead is of utmost importance, and your llama should be trained to accept and stand quietly on a picket line. A picket line is a stake that is screwed into the ground (dog stakes can be used, and are available at pet and hardware stores) with a long line attached to allow the llama to browse and graze freely.

You can train a llama to accept a picket line fairly easily. Stake out your llama at home for several minutes a day, and stay with him. Be sure he does not become entangled in the line. He will quickly learn to untangle himself without a fuss by simply lifting his leg and allowing the line to fall off. Soon your llama will be grazing quietly while he is tied to a picket line.

If possible, begin with a yearling. By this age, llamas are mentally mature enough to accept training. A yearling probably has already been taught to halter and lead, and perhaps you have already worked him over small obstacles. He is eager to learn and probably has not developed a mind of his own. Trust is critical and this is an ideal time to develop it.

You might bring your yearling along on a hike with veteran pack llamas, "caboose"-style. Llamas are extraordinarily easy to train, but they learn even more quickly by observing and imitating the actions of other llamas.

Conditioning and Weight Training

Conditioning is a word with several meanings. Conditioning can refer to the fitness of the llama, as an athlete that has trained properly and is in top physical form. Another form of conditioning refers to proper weight. In this case, underconditioned means that a llama is too thin; overconditioned is a polite term for a fat llama.

The best place to check the conditioning of your llama is at the backbone. Run your hand down the llama's back, keeping your thumb on one side of the backbone and your other fingers on the other. You should be able to feel the animal's backbone but not fatty deposits along the spine. (See Chapter 3 for information on body scoring.) If your llama is underconditioned, you will be able to feel individual vertebrae. Also check the rib cage. You should be able to feel the animal's individual ribs, but they should not be so pronounced that he feels bony. Have someone lead your llama away from you. Look at his hindquarters. Does he possess a nice U shape between the legs, or are his thighs touching each other? (Examining your llama in this manner applies not only to a pack llama, but also to any llama whose weight you wish to check.)

Weight training is an integral part of preparing llamas for packing. Begin with a soft pack, or even a blanket folded up on his back. Llamas are not bucking broncos; they will readily accept the light pack if they accept brushing and grooming. Start over a mile of moderate terrain with 10 to 15 pounds. (Llamas can appear to be fully grown at 2 years of age, but their bones, joints, and tendons are not yet fully developed.) *Until the age of 3, any weight over 30 pounds is not advisable.*

Just like athletes, llamas must be in good physical condition, and the weight they carry should be increased on a gradual basis. With each addition of weight, add more length to his trip. Take notice of your llama's attitude: Is he still willing after several miles? Is he beginning to slow down? Does he have difficulty catching his breath? These are important signs. They tell you when he has had enough for the day. Give him a day's rest and then repeat the process, adding 1 mile and 10 pounds each trip. *On the average, working three days each week for a month, and achieving 10 miles a day by the end of the period, will guarantee you a fit and conditioned packer who is ready to work.*

Dennis Jensen

A mature llama that has been properly weight trained and is physically fit can carry one-third of its body weight up to 12 miles per day, depending on the terrain and the temperature.

Consider the kinds of trails and trips you plan to undertake. Be sure to train your llama under a variety of conditions. Take him on short trips on a rocky trail, or over hilly terrain. Try him at different altitudes. Work with him in the rain. On a long trip, you may encounter a range of conditions. Both you and your llama must be prepared for them all.

Saddles and Tack

There are numerous manufacturers of llama pack saddles and related equipment. Because of the shape of a llama's back, the animal requires a saddle that is specifically designed for llamas. Llama pack equipment is especially lightweight. Many manufacturers will accommodate a new llama packer by offering a return policy should the saddle and pannier system not fit properly.

Be sure to use a saddle pad beneath rigid pack saddles.

Saddles

A llama pack saddle is fastened by one or more girths to the llama's back. This is a foundation "saddle" to which the panniers are attached. (Panniers are packs that llamas carry on their backs. They are similar to saddlebags.) Llama panniers are modified to fit the llama's pack saddle. There are two types of pack saddles. One is a rigid frame saddle, generally made from wood, aluminum, or fiberglass. The frame distributes the weight on either side of the llama's spine. The pack frame is used with a saddle pad.

A rigid frame saddle with a crossbuck design will support more weight.

This basket pack is handsome, but it is less practical than panniers made of Cordura, canvas, or similar material.

A soft pack protects the back and spine, and distributes weight evenly on each side.

The second type of saddle is a soft pack. It is constructed from leather or nylon and includes some type of padding to protect the back and spine. Here, too, the weight is distributed on either side of the llama's spine.

The type of packing you'll be doing will determine which type of system you use. When you try a saddle on your llama, be sure it fits comfortably and does not dig into his flesh or pull on his wool. The saddle should never rest directly on the llama's spine, nor should it come directly in contact with the spine. It should readily clear the spine, allow free shoulder action, and distribute the weight evenly across the animal's back. Llamas come in different shapes. Try both types of pack systems by different manufacturers to ensure a comfortable, proper fit.

Britchings

Depending on your llama's physical structure and the type of packing you will do, you may need to add additional strapping to the standard two-cinch girth. If you are packing in a mountainous area with steep inclines, you will need to utilize britchings. *Britchings* consist of a breast collar similar to a pack mule's or horse's and a crupper that straddles the llama's hindquarters and fits under the tail. The *breast collar* stops the saddle frame from sliding backward on steep inclines. The *crupper* prevents the saddle from sliding forward on steep descents.

Panniers

Saddlebags for a llama packing system are called *panniers*. Most are made to accommodate the commercial pack saddles currently on the market.

Kevin Kennefick

Manufacturers offer many different types and sizes. Be careful not to choose panniers that are too large for your llama. In the rigid pack system, the panniers hang from the frame's crossbuck design. You don't want the panniers to interfere with your llama's freedom of leg movement. Generally, soft packs come with their own set of panniers.

Panniers that are manufactured to accommodate a soft pack saddle system are not interchangeable, as they fasten differently to the soft saddle. Rigid pack saddles made from aluminum or wood are normally of crossbuck design, and the panniers loop over the crossbuck or saddle horn for fastening.

Be sure the panniers fit your llama properly; they should not interfere with his leg movement.

Weighing Panniers

With either system, at times you'll have to weigh the panniers and their contents. Each loaded pannier should be within 1 to 2 pounds of the other. Uneven weight distribution can wreak havoc on the stability of your saddle and load, and cause discomfort for your llama. An unbalanced load will slow your llama's pace and unbalance him on steep or uneven terrain.

Hitting the Trail

A pack llama is a work llama. Shear him in the spring before you start his conditioning. A shorn coat is easy to care for, will not collect debris, and makes it easier to saddle your llama for the trail. Moreover, a pack llama that has been shorn is more comfortable working in warm weather.

Transporting Pack Llamas

You're ready to head for the trailhead. Llamas are adaptable and can be transported in the bed of a pickup truck equipped with stock racks, a van, a minivan, or a horse trailer. You don't need to tie a llama when you transport him in an enclosed trailer or vehicle, but it is advisable to tie him in a pickup or open trailer. Use a quick-release knot or a breakaway snap; be sure you can free the llama easily, if needed. A llama may jump out of an open vehicle if he is not tied, but make sure he has enough line to permit him to kush (lie down).

Preparation

To prevent any discomfort to your llama, brush all debris from the fleece on his back before you saddle him up. Make sure all straps and cinches are over the top of the saddle so they don't interfere with his legs. Now place the saddle on the pad, keeping it just behind his shoulder blades. Fasten the front cinch first, as tightly as possible, followed by the rear cinch. The rear cinch should be approximately 4 inches in front of your llama's sheath (the penile area). You should be able to push your fingers through both cinches; otherwise, they could restrict your llama's breathing.

If you are using a breast collar, you should be able to slide your hand between it and the llama's chest. Be sure that it is not hanging where it could interfere with your llama's leg movement. If you are using a crupper, your hand should be able to slide between the crupper and the llama's body. If you are using an intact male, be sure that the crupper is not so high that it will rub his testicles and make him uncomfortable.

Just behind the midshoulder is the llama's center of gravity. The front of your llama will carry 60–65 percent of the pack load. The pack saddle should fall just forward of the center of your llama's back, and the heaviest load in the panniers should be about one-third of the way from the top of the load.

Packing the Panniers

Properly packed panniers are easier for your llama to carry. Compact, tight loads are also easier to arrange for weight distribution. A hand scale can be purchased (for about $15) to weigh the panniers. Pack the panniers as evenly as possible. Place the larger, heavier items in the panniers first, then tuck in the smaller ones. Careful packing may sound like a chore, but with experience you will be able to look at an item and know immediately how much it weighs and where to pack it.

Remembering the llama's center of gravity, try to keep your loaded panniers as close to the llama's body as possible. Panniers that are not fastened properly will flop around when the llama moves and will tire him out. Panniers that are too large and hang below the llama's belly create additional strain. A llama does not possess an even, four-beat gait like that of a horse; a llama uses a pacing gait. A pacing gait means that both feet on the same side move forward simultaneously, creating a side-to-side gait. This gait allows the loaded panniers to shift more easily, so keep the top of the loads lighter — perhaps sleeping bags or blankets — as anything heavier will end up sliding off to one side.

Nutritional Requirements on the Trail

Working pack llamas are defined as llamas on the trail a minimum of 4 to 6 hours per trip, three times a week. These llamas need a minimum of 10 percent protein in their diet, but not more than 12 percent. If your working packer is on this type of increased protein diet, then he will not normally need you to pack grain for him. Browsing and grazing should be sufficient. However, grain will need to be packed if he is working excessively hard, as in extreme heat, with extra heavy loads, or along continued steep or rough terrain.

First Aid Supplies for the Trail

- vet wrap
- gauze or rolled cotton
- iodine solution (7%)
- ophthalmic ointment
- duct tape
- toenail clippers

The conditioning of the llama will dictate how much rest he needs. As a rule, if your llama is carrying 10 pounds or more than he is usually used to, he will need a rest of about 15 minutes every several miles. In temperatures above 80°F, he will need to stop every mile. Terrain also plays a part in how often to rest your llama. A good indication is if your llama is beginning to slow, breathe heavily, or balk. He is telling you he is tired.

Gale Birutta

Yin's appetite on the trail includes such tasty morsels as Black-Eyed Susans.

Offer your llama water on these breaks, especially in hot weather. You may need to carry water if you will be packing away from streams during bouts of hot weather.

Because llamas are browsers and like to nibble at this and that, unless you are on an extended trip for several days where your packers will be working hard, you won't need to bring additional grain. Llamas should be offered at least one long, good drink per day. Llamas will drink 1 to 3 gallons of water daily — perhaps at a moving stream. Don't be alarmed if they don't imbibe: You can lead a llama to water but you can't make him drink!

Day's End for Your Llama

After an enjoyable day, don't forget that your llama friend has worked hard to please you and your guests. Make him comfortable, and by all means enlist your guests for assistance. They will be honored that you asked and more than eager to help.

Whether you are back at the barn or still on the trail on a several-day outing, at day's end promptly unpack the llama and remove his saddle. Check his feet and toenails for any signs of wear or injury. Brush off his pack and offer him a drink of water. If you are out on the trail, tether him where he has access to water and forage. If you are at home, turn him into the field with his buddies with fresh water.

Managing a Successful Business

Target Your Trips

Decide what kinds of hikes you will offer and then cater to that market. For example, do you want to offer easy-going, family hikes? Get in touch with your local chamber of commerce — it is an excellent source of promotion for any type of pack trip, and one of the first places family vacationers will contact.

Are you more interested in appealing to avid outdoorspeople? Advertise in outdoors or sports magazines. And again, your chamber of commerce will have marketing ideas.

You'll notice that I refer to hikes and trips. I avoid the word "trek" because it may be misleading. In horse trekking, people ride horses. With llama packing, you *walk* along with the llama. I stick to "hike," "trip," and "walking tour."

Gale Birutta

Leisurely half-day hikes are popular with couples and families.

Regardless of the type of people you want to target, consider offering several types of trips. This way you can cater to a broader range of clientele. For instance, if you have chosen lightweight packing, offer both half-day and full-day trips. Plan your route on a variety of terrains, accommodating both families and more proficient hikers.

Establishing Fees

You may find it difficult to establish a schedule of fees. The local economy will have an effect on what you charge, as will the type of trip and whether you will provide meals. Write for other llama packers' brochures in your area and see what they charge. If you can't find any others for comparison, a good guideline is to start at $25 per person for a half day with a light snack. If you add lunch, increase the charge by $10 per person. Set a full-day hike at a minimum of $50 per person.

I charge more per day than any other llama outfitter in New England ($45 per person for a half day with light lunch; $95 per person for a full day with a gourmet meal prepared on the trail). I have been offering trips for years, and my reputation draws a serious clientele.

Depending on the terrain in your area, half-day trips will cover from 3 to 5 miles in 4 hours. Full-day trips take about 7 hours and cover 7 to 10 miles.

Insurance

Commercial outfitter's insurance is expensive. Check with the company that carries your farm liability. For as little as $100, many will add a rider for $1,000,000 in umbrella liability to cover your packing operation. If an insurance company classifies llamas in a high-risk category with horses, *look for another carrier.* These horse rates will shut you down before you even start.

Promotional Literature

People want to know what they are spending their money on, and a good promotional brochure will provide them with this information. It should be informative and descriptive. Be thorough: Inform the customer about the length of the trips (hours and miles), the scenery and terrain, points of interest, and meals. A color brochure full of pictures would be nice, but also very expensive. If the writing is lively and you include a few pictures, you won't need anything flashy. But if you can afford some color, think about incorporating your farm colors.

Generating Customers

You now have trained packers and all your equipment — all you need now is customers. Here are some useful suggestions on how to start or increase your business.

Contact journalists and invite them — along with their photographers — on a trip. (Journalists are not permitted to accept a no-charge offering, so they will pay your fee.) Organize a mailing to major newspapers in your state and of course to the local papers. Include national newspapers, such as the *Boston Globe* and the *New York Times*. You'll get a tremendous response for in-the-field (on pack trips) interviews, and your only major expenses will be the cost of the mailing and your time. And photographers will appreciate not having to carry their equipment.

Newspaper journalists are accustomed to deadlines, and their stories will most likely appear within a week of reviewing the trip. A feature with pictures will generate an instant flow of business. There is no better exposure for your trips than a newspaper or magazine article.

Booking Trips

Booking trips can be handled in several ways. Ideally, you'll want to accept charge cards. If your business does not warrant charge cards, then accept telephone reservations, advising the customer that you can hold a date for only five days. Require a 50 percent deposit per person, by check or money order, to formally book that date. (The check should arrive with enough time for you to make sure it doesn't bounce.) This will get the customer to act immediately and will also weed out those who are not sure if this is what they really want. If you have half the money, they will probably show up — if not, you lose nothing.

Requiring the deposit also creates the impression that trips are in demand and encourages a speedy decision to reserve. Use a reservation form that also acts as a release. Remember, a release does not mean you won't be sued — *carry insurance*. (Sample reservation and release forms are on pages 242 and 243.)

Require that the balance be paid *prior* to the trip departure. It makes life easier and you are assured of payment even if someone did not enjoy his trip.

> **Tip**
>
> *Never* book a trip without a reservation deposit, or you risk standing alone at the trailhead without clients.

Personalize Your Trips

During the packing season, concentrate on personal service. Try offering "custom" adventures, meaning that you are willing to accommodate each group specially. To keep a good profit margin, establish a policy that all trips must have at least four people. Usually a couple will find some friends to join them in order to meet the group minimum. As for "custom," each trip is based on the group's hiking or walking capabilities.

A particular favorite custom trip I once led was for a group of wild mushroom enthusiasts. This trip was specially planned as I had to seek out a wild mushroom specialist and depend on her knowledge to point me in the right direction on the trails where certain wild mushrooms could be found. The specialist accompanied the group, and I did enjoy learning about edible wild mushrooms. The group paid for the specialist and the pack trip itself.

Another trip was an outing for a women's group that wanted to learn all about llamas over a three-day period. The first day started out with a general video and discussion about llamas at the bed and breakfast where they stayed. The second day consisted of a full-day pack trip, while the third day was a hands-on experience where these gals caught, haltered, groomed, and tacked up the llamas for the remaining half-day trip. Many of these women have returned year after year.

Food and Menu Planning

Food is a very important factor on pack trips. The fancier the food, the happier the hiker, especially if he is paying a good price. On multiday trips, your guests expect three square meals a day. Some who book a trip are looking at the meals as a major component of the llama packing experience. Check with your local Board of Health regarding licensing and health regulations. Do not allow anyone to bring his own food. You are the only one who knows how to pack llamas correctly and what foods can be packaged appropriately.

Gale Birutta

Meals and snacks that are carefully planned and attractively presented will keep your customers coming back.

Roll-up tables that can be carried on the crossbuck of the saddle are available from suppliers.

Keep the menu simple yet imaginative. Meals should be arranged to suit the customers. Consider the people you are catering to: Picnic-type lunches are good for family groups; hearty meals will satisfy the hungry seasoned hiker. If any of your customers are vegetarians, you will know from their reservation forms. Make sure you accommodate their needs.

Your guests' metabolism speeds up while hiking, so think about added carbohydrates. Rices, breads, pasta, and legumes are excellent additions to any trail meal. Make pasta sauces at home with fresh ingredients, freeze, then reheat at mealtime. For day trips, lunch can be as simple as pita bread filled with vegetables, salads, cheese, and meat. (Sample recipes are on pages 240 and 241.)

If you are accommodating families, remember that some children are picky eaters. Ask the parents about food preferences. This might mean preparing and serving a separate meal for kids, so consider your profit margin and the time you are willing to spend.

In lightweight or heavyweight packing, make the meal a big event. A fabric tablecloth and matching napkins will dress up your meals. Depending on the season, try a fresh flower centerpiece. On any type of trip, formal buffet-style is the most convenient. This works best when a spouse, friend, or employee can entertain your guests while you prepare the meal. The buffet is set up while a guide takes the group on a short hike to explore the area and its interesting landmarks or

Dennis Jensen

views. The guests will be awestruck by the buffet when they return. Allow at least 90 minutes for a meal. You want guests to relax, to chat, and to enjoy the food at an unhurried, leisurely pace. With their tummies full, they'll soon be eager to resume the hike.

Tips for Packing Food

Coolers will keep eggs fresh for several days. If you are planning scrambled eggs or using eggs in other recipes, break them open into a large container and mix them before you leave. (Eggs also may be frozen.) You will want to use fresh foods, but be sure to prepare as much as possible at home before departure. Do all your chopping, grating, marinating, and cooking of sauces at home. Tupperware or similar plastic containers are lightweight and come in assorted sizes. Ziploc freezer bags are also excellent as they can be used from original preparation to freezing.

Diversify and Enjoy

The more you offer during the day, the better. Be bold — don't be afraid to try new approaches. Both family groups and avid outdoorspersons will enjoy a fishing trip. The llamas can carry inexpensive, inflatable boats that will take you to those hard-to-reach boiling spots. (Boiling spots are areas that are filled with fish. In the early morning or early evening, the pond or river will appear to "boil" — like boiling water — with fish in a feeding frenzy.) You can find other uses for these boats: It's a nice break on a hot day to paddle around on a serene pond or lake. Your llamas can carry inner tubes as well. Your customers will love fishing or floating, and your llamas will enjoy a well-earned rest.

Word of mouth is wonderful advertising — and it's free! If your customers leave delighted with the day, they will tell all of their friends. Many of your customers will return every year, bringing a new group of friends with them each time. Loyal customers may become attached to a particular llama and may even ask for him by name.

Llama packing is a marvelous experience. When you target the right market for the trips you offer, you'll enjoy profitable and fun-filled days with new friends. You'll look back on every adventure with fond memories, recalling details that made each one special. And you'll have been well paid, too.

Venison Vermont* — The Favorite!

This tasty dish uses less sugar, salt, and fat. Recipe includes *Diabetic Exchanges*. Can be prepared at home and just grilled on the trail.

        1   cup cooking oil
        ⅔   cup cider vinegar
        2   tablespoons Worcestershire sauce
        ½   medium onion, finely chopped
        ½   teaspoon salt
        ½   teaspoon sugar
        ½   teaspoon dried basil
        ½   teaspoon dried marjoram
        ½   teaspoon dried rosemary
    2 ½   pounds boneless venison (or lean pork, beef, lamb, or chicken),
             cut into 1½- to 2-inch cubes
           Italian rolls *or* hot dog buns

In a glass or plastic bowl, combine first nine ingredients. Add meat and toss to coat. Cover and let marinate for 24 hours, stirring occasionally. When ready to cook, thread meat on metal skewers and grill over hot coals until meat reaches desired doneness, about 10–15 minutes. Remove meat from skewers and serve.
* May substitute pork, beef, lamb, or chicken.

Cheese and Potato Wild Rice Soup

Make this ahead of time and carry in a thermos. Excellent for crisp autumn days.

        ½   cup wild rice, uncooked
     1½   cups water
        ½   pound bacon, cut into pieces
        ¼   cup chopped onion
      Two   10¾-ounce cans cream of potato soup
             (dilute with 1 can liquid — ½ milk; ½ water)
        1   quart milk
     2½   cups grated American cheese
           Carrot curls (optional)

Combine wild rice and water in saucepan and cook over low heat for 45 minutes. Drain. Set aside. Fry bacon pieces and onion in skillet until bacon is crisp. Drain bacon and onion on paper towel. Place soup in large saucepan; dilute as directed above. Stir in milk (1 quart), bacon, onion, cheese, and cooked rice. Stir until cheese is melted. Garnish with carrot.

Pound Cake Poppy Seed Muffins

These are very filling muffins. Bake ahead and serve at room (or outdoor) temperature.

> 2 cups all-purpose flour
> 1 tablespoon poppy seeds
> ½ teaspoon salt
> ¼ teaspon baking soda
> 1 cup sugar
> ½ cup butter or margarine
> 2 eggs
> 1 cup plain yogurt
> 1 teaspoon vanilla extract

In a small mixing bowl, stir together flour, poppy seeds, salt, and baking soda. In large mixing bowl, cream sugar and butter. Beat in eggs one at a time. Add yogurt and vanilla extract; mix well. Stir in flour mixture until dry ingredients are moistened. Spoon batter into greased muffin tins. Bake at 400°F for 15–20 minutes or until a wooden pick inserted in center of muffin comes out clean. Cool muffins on wire rack 5 minutes before removing from pan. Yield: 12 muffins

Spiced Chinese Chicken Wings

This is a favorite with kids. Make it ahead and serve cold.

> 12 whole chicken wings
>
> Sauce: ¾ cup soy sauce
> 1 clove garlic, pressed
> ½ teaspoon dark roasted sesame oil
> ½ teaspoon powdered ginger
> Pinch of Chinese 5-spice powder

Remove tips from chicken wings. Cut each wing in half at joint; set aside. Combine sauce ingredients in bowl or heavy-duty plastic bag; add wing pieces and marinate, refrigerated, 1 hour or more. Remove wing pieces from sauce and place, thick-skin side down, in a lightly greased, shallow baking pan. Pour sauce over wings. Bake at 375°F for 20 minutes. Remove from oven and turn wings. Return to oven; bake 20 minutes more until browned. Serve hot or cold.

Treats au Llama

- Sliced carrots or zucchini
- Sliced apples
- Small amounts of horse sweet feed

Reservation Form

Please complete this reservation form and submit it with your deposit.

Minimum deposit per person is 50 percent. Balance due day of trip prior to departure. Make checks payable to Llama Expeditions.

Please hold _____ reservations for the following llama pack trips:

(type or name of trip) _____

(date/dates) _____

Enclosed is $ _____ deposit. Vegetarian? Y _____ N _____

Our balance of $ _____ will be paid day of trip prior to departure.

Contact Person:

Full name _____ Age _____

Mailing address _____

City _____ State _____ ZIP _____

Home phone ( ) _____ Business phone ( ) _____

Other members of party:

Full name _____ Age _____

Mailing address _____

Full name _____ Age _____

Mailing address _____

Full name _____ Age _____

Mailing address _____

Full name _____ Age _____

Mailing address _____

Full name _____ Age _____

Mailing address _____

Use additional sheet if necessary

I have read, understand, and accept the terms, conditions, and policies of Woolly Llama Expeditions.

_____ Date_____

(contact person signature)

_____ Date_____

Release Form

Eligiblity: We require that everyone who takes our tours be in reasonably good health. Children under 18 must be accompanied by an adult.

Meals/snacks: We serve only home-cooked meals and snacks using fresh ingredients. We offer vegetarian dishes for all trips, in addition to our regular menus. Please advise us of any special dietary needs when you receive your reservation confirmation. NO alcoholic beverages are permitted.

What to bring: The best part of traveling with llamas is that they carry all the supplies — no struggling with a backpack or shoulder carrier. Please keep in mind, however, that we still need to travel light. We limit each traveler to 15 pounds unless special arrangements are made beforehand. A detailed checklist is enclosed.

Equipment: Woolly Llama Expeditions will supply all equipment, including eating utensils, tables, and camp chairs.

Transportation: We will forward to you under separate cover detailed directions to our home trailhead. In case of departure from a trailhead other than our own, you will receive those directions. In this case, do your best to car pool to the trailhead.

Reservations: As our trips and walking tours fill up quickly, we suggest that you book your trip well in advance. Telephone reservations are accepted; however, they will only be held for five days, at which time we must have a 50 percent deposit to book that date formally. Reservations are confirmed on a first-come, first-served basis. If a date is full, upon receipt of a 50 percent deposit, we will place you on a waiting list.

Payment: A 50 percent deposit per person is required at time of booking. Should you need to cancel, we can reschedule your trip. However, deposits are refunded only when your original date is filled. The balance of the payment is due on the day prior to the departure of your trip. In the case of cancellation because of weather, your trip will be rescheduled or we will refund your deposit.

If you must cancel your reservation, any refund will be based on the following:
1. 30–45 days or more prior to departure: full amount less $25 per group handling fee;
2. 20–29 days prior to departure: full amount less $35 per group handling fee.
3. 1–19 days prior to departure: No refund unless date is filled from a waiting list.

We reserve the right to cancel any trip at our discretion to ensure the well-being and safety of our guests. We shall notify everyone with reservations as early as possible. In this case, we shall reschedule your trip or refund your deposit in its entirety. Anyone canceling from a waiting list shall also receive a full refund. We reserve the right to use or assign the use of any photographs taken by us of any part of our backcountry trips for private, commercial, educational, and promotional purposes.

Emergency care: Every precaution is taken to ensure each person's safety and the careful handling of your belongings. We utilize only the safest route for our trips and we are fully insured. We shall not assume any liability whatsoever for any personal injuries or loss or damage to personal property in the absence of negligence on our part. Medical care is the financial responsibility of the ill or injured person. Professional medical assistance is not offered on our trips.

Native foliage: Please preserve our natural environment. Please refrain from picking any type of foliage or wildflowers. Only in designated areas may you enjoy the picking of blueberries, raspberries, or blackberries. Your guides will advise you.

Miscellaneous: At no time are firearms of any type permitted on our trips with the exception of drop camps or hunting expeditions. Pets are not permitted. Llamas are strictly pack stock and are never ridden. Parents must be able to carry children who tire.

THIRTEEN

FLEECE

Fiber Glossary

Fleece: Unwashed and shorn fiber from a llama or alpaca.
Fiber: The unshorn haircoat on a llama's or alpaca's body.
Wool: A generic term for fleece or fiber.
Crimp: The natural curl grown into the fiber from the follicle.
Curl: Curl is similar to crimp, but generally straighter.

Judith Chorney

Llama fiber has been a valuable product for thousands of years. But it is only recently that breeders in North America have been taking a serious look at llamas as producers of fiber.

Llama fiber has the same uses as any other natural fiber. Clean, lanolin-free, light, and odorless, llama wool is a favorite among hand spinners. Fine spun yarn can be processed into fabric yardage. The white wool takes dye well. Llama wool is coarser than that of alpacas, but owners can breed for finer wool.

You can make lovely sweaters, hats, mittens, and scarves from llama wool.

Llamas as a Fiber Producer

Although llamas were originally bred as beasts of burden, their owners valued their fiber to make warm clothing in unforgiving climates. The wool was used to make ropes and blankets as well.

Llamas and alpacas offer a broad range of natural colors from white and beige to light and dark brown, to grays, blacks, and reds.

History

Andean llamas were domesticated over 5,000 years ago. From the 11th to the 13th century, the Incas practiced sophisticated herd management and selectively bred for fleece quality and strength. The conquering Spaniards destroyed the Incan civilization and decimated the llama population. Even today, the Incas' herd management practices rank among the top accomplishments in animal husbandry. (It has only been in the past century that any wildlife management practices have been in effect in the modern world.)

Today South Americans are once again managing their herds to improve both the animals and the quality of the fleece. North American alpaca and llama breeders are also improving camelid fiber through proper diet and selective breeding, with dramatic success.

Today in North America, llamas are high-quality fiber producers. With the shift in the mid-1980s to an interest in natural fibers, fiber producers began looking for alternatives and additions to sheep's wool. Many sheep producers discovered that llama fiber blends well with the wool of sheep.

Because there is so much interest in llama fiber, it is possible to find workshops on the care, use, and grading of llama fiber. The International Llama Association (ILA) has worked hard and publicly for about 15 years to promote llama fiber. Regional associations tout the fiber through fairs, shows, and bazaars.

Recently the trend has been to shear rather than brush the fleece. Repeated brushing seems to damage the llama's coat. Shearing llamas every year or every other year creates a healthier fiber. Shearing has other benefits as well. In warm climates, animals are cooler during the hot seasons and thus more comfortable. Breeding animals can perform better because they suffer less from heat stress.

Shearing is also a good time to trim hooves, check udders, worm all the llamas, treat them for ticks and parasites (after the wool is removed), and check the teeth of older llamas.

Physical Characteristics

Generically, llama fiber is "wool." But llama fiber is a medullated fiber (technically considered hair because of its hollow core), and therefore not really wool. Sheep's wool is a solid fiber. For the purposes of this chapter, though, I will use the word "wool."

Llama wool presents a 90 percent yield after washing. The yield is so high because the wool contains only about 10 percent lanolin. (Lanolin is a grease usually found in wool. Repeated washing removes lanolin.) In comparison, sheep's wool has a 40 to 50 percent yield because of lanolin and debris such as manure and hay. Llamas, with their looser, more free-flowing, and greaseless fleece, generally stay cleaner than sheep. Llama wool is lightweight because of its medullar structure. Its hollow core provides remarkable warmth and outstanding insulating capabilities without bulk. Because of its tensile strength and durability, llama wool also shrinks little during washing or processing. And being hypoallergenic, llama wool can be worn by infants and others who can't wear sheep's wool.

Llama fiber yields a yarn that is lightweight, warm, and 90 percent lanolin-free.

Llama yarn is a consistent winner in fiber and hand-spun yarn categories at farm shows and local, county, and state fairs.

Llama wool is similar to sheep's wool in the surface of irregular overlapping "scales." Scales are the structures that hold the fiber together. Llama wool has fewer scales, resulting in a softer, finer fiber. But because it has fewer scales, llama wool lacks the elasticity of sheep's wool. A blend of 80 percent llama wool with about 20 percent sheep's wool provides good elasticity.

The llama is a two-coated animal. The outer coat of crimpless guard hair acts as a moisture barrier, repelling rain and snow. The downy undercoat is the insulator that provides warmth against the elements. This downy undercoat has more crimp and more elasticity than the outer coat. The excellent diet and selective breeding practices in North America have improved the wool of many llamas. Some llamas exhibit an increase in fineness and length that rivals even alpaca fiber.

Alpaca Fiber

Unlike llamas, most alpacas are solid in color. There are 22 color variations derived from 7 basic colors: white, light fawn x, light fawn z, light coffee (or brown), medium rose gray, medium silver gray, and black. Alpacas are single-coated animals, and their fleece lacks guard hairs. Alpacas supply 5 to 8 pounds of fiber per shearing. Strong and resilient, alpaca fiber — regardless of its fineness — retains its strength, making it perfect for commercial processing.

Genetic research in Australia, the United States, and Canada focuses on the improvement of the alpaca and its fleece. In the past few years, alpacas have, through selective breeding, experienced an increase in fleece weight per animal and quality of fineness of the fiber.

Productivity

Several factors affect fleece quality: climate, age, breeding activity, and diet. Female llamas will produce healthy fleeces until they reach breeding age. At that time, hormonal changes occur and energy is diverted to pregnancy. Male llamas produce healthy fleeces until about 12 years of age. At that time, hormonal changes begin and the fleece naturally starts to deteriorate.

Fleeces generally grow at the rate of 4 inches per year. An adult coat yields 3 to 8 pounds of fiber yearly. Shearing encourages the highest yield.

Brushing is time consuming and generally makes the yield less. Brushing does have one advantage,

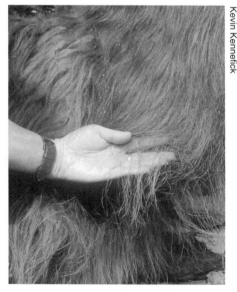

Kevin Kennefick

Guard hairs protect the downy undercoat of the llama.

however — only the downy undercoat of the llama is harvested. The outer guard hairs remain as a protective coat. In contrast, during shearing, both the downy undercoat and the guard hairs are harvested. You must then separate the guard hairs at the time of shearing or pay for it to be done commercially by a mill.

Fleece Evaluation

Wool is measured in microns. A micron is $\frac{1}{25,000}$ of an inch or $\frac{1}{1,000}$ of a millimeter. Alpaca wool averages 22 to 28 microns; llama averages 25 to 30 microns. Human hair, in comparison, measures 90 to 100 microns. There are several labs that will test the micron diameter of your fiber. (Contact the Alpaca Owners and Breeders Association, listed in Appendix A, for a lab referral.)

Micron diameter alone is not the only test of the value of fiber, however. Consider the following when you evaluate the quality of your llama's fleece:

Coverage. Coverage does not always mean quality in North American llamas. Coverage refers to the amount of *continuously growing fiber* on the overall body surface of an animal. (The entire fleece coat of a llama does not grow all at the same time. Fleece on the legs, head, and neck can grow very slowly. The part of the fleece that does continue to grow is known as continuously growing fiber.) An animal may have very high quality fleece with poor coverage. Gently open the fleece to the roots in several places to discover where the highest quality fiber grows on your llama.

Crimp/curl. Crimp is the waviness of the fiber's length throughout the blanket of the fleece. Crimp is a highly desirable characteristic and will determine the quality of the manufactured product. A fleece with more crimp holds its shape well. Excessive grooming removes the crimp.

Crimp is what makes the fleece elastic or stretchy. Fleece with good crimp acts like a spring. When fleece fibers are stretched, they spring back to their original shape. As a rule, the finer the fleece strands or fibers, the finer the crimp. In very fine fleece, the crimp can barely be seen. In very coarse fleece, the crimp can vary from $\frac{1}{2}$ to 3 inches between the "zip" and the "zag."

The llama's body is protected by its downy undercoat.

Kevin Kennefick

Curl, which is straighter than crimp, has a more hairlike quality. However, curl can occur with crimp. Examine curl by gently opening the blanket of the fleece.

Fineness. Individual fibers are measured in microns (see above). Remove a piece of the fleece and evenly spread out the individual fibers. You will be able to see the diameters of the fibers and how consistent they are within the particular piece.

Density. Density can be measured by determining how many fibers a particular llama grows per square unit area of skin. The ability to determine density comes with experience. You may get the hang of it from practicing on only a few llamas. Part the animal's fleece and examine closely the amount of skin exposed at the fiber's roots. Do the same with two or three more llamas. You will be able to pick the animal with the densest coat by comparing it with the others.

You may want to attend a workshop on llama fiber and its evaluation. The knowledge you'll gain from these workshops is well worth the time spent.

Harvesting Fiber

You can collect and harvest fleece by brushing or by shearing. When you remove fleece, it is important to maintain the marketability and visual appeal of the llama. Even if your only interest is in fiber, eventually you will want to breed your animals for their fleece. They'll have to look good to attract breeders.

Brushing

Brushing is a way to harvest fiber from your llamas, and if you do it in moderation, it promotes healthy skin and aids in circulation. There are several styles of brushing, from basic surface grooming to grooming "to the skin." The latter is not recommended because it damages the fiber. Reserve this type of thorough grooming for exhibitions, shows, and sales. Excessive grooming removes some of the crimp or curl, and as with human hair, excessive brushing will cause split ends and frizziness. Detangling products will condition the coat and alleviate much of the stress the animal experiences from long periods of brushing.

In surface brushing, you brush just the top coat, freeing the fleece of mats and tangles on the surface. You won't be able to harvest any salvageable fiber from surface brushing.

The next stage of brushing takes in additional layers. This enables you to get into the undercoat of the llama, where the best fleece is found. Depending on how much the llama will tolerate and the density and condition of the fleece, you can easily come up with 2 or 3 ounces of fiber in a short time, especially if you brush during shedding.

The Natural Shedding Process

Llamas naturally shed all year long, as the downy undercoat is constantly replacing and regenerating itself. Shedding is more evident in the spring, however. Llamas enjoy being brushed to remove shedded hair, although it is not a necessity. The undercoat will naturally work itself out and fall to the ground.

It takes approximately two years for a llama's fleece to reach its full length and density, and that's when he will start shedding. Usually, you'll notice shedding (normally in spring) first in the neck and leg wool areas. Shedding occurs every other year in short- to medium-wooled llamas. Three or 4 inches of undercoat is the average length on a light- or medium-wooled llama.

The ability to shed has for the most part been bred out of heavily wooled llamas. (Heavy-wooled llamas are those with 5 or more inches of undercoat.) These llamas are most susceptible to hyperthermia, or overheating, and can experience fertility problems and even death in extreme cases. But cases of hyperthermia usually are only reported in very heavy, *unshorn* llamas. These particular llamas may also be heavily matted. It may seem that the industry is breeding for a weakness, but with today's better management practices and different types of shearing cuts, this is not the case.

Brushing is one way to harvest llama fiber.

A wide-toothed comb can be used to harvest llama fleece.

Shearing

In the 1980s, it was rare to see a shorn llama. With the emphasis on heavily fleeced animals, the practice was taboo. By 1990, good herd management dictated that most heavily fleeced llamas be shorn for health reasons, primarily to alleviate heat stress. Reproductive problems can result in the unshorn, heavily fleeced male during overly hot weather; in particular, impotence and sterility.

For fleece harvesting, shearing remains the preferred technique. Guard hair remains with the fleece if the fleece is sent to a commercial mill for processing. In processing, the guard hair will float to the top during washing. If you don't send the fleece out, you can remove the loose guard hairs yourself. Grab with both hands the longer, cut guard hairs and pull them gently from the fleece.

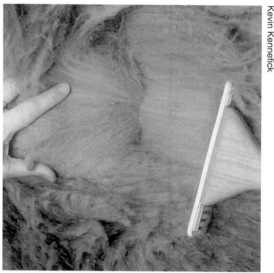

The undercoat of the llama is where the best fleece is found.

Why Shear?

Regular trims stimulate growth of healthy fleece. For breeding animals as well as work llamas, shearing is a good practice for comfort and overall health. Most of the sweat glands are in a llama's armpits, groin area, and belly. Here the skin is the thinnest, allowing the body to cool. A heavily

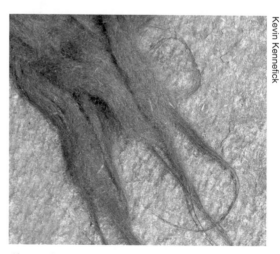

If you shear your llama, you'll need to separate the guard hair from the more valuable undercoat.

wooled llama will have trouble cooling off when heavy fleece covers these areas. The cooling process stops almost entirely, even when the llama positions himself in the "tea-cozy" position.

With a breeding stud, these crucial cooling areas are covered while he is copulating. If your sire is nervous and continually paces to guard his territory, shearing is even more important because he can become overheated from excessive or continual physical activity. Most heat loss is from the llama's legs, neck, head, and hind end. This may explain why llamas tend to lose their head, neck, and leg wool first.

You'll be able to treat skin problems more easily when matted fleece is gone. Air and sunshine will help the healing process. A matted fleece traps humidity and hinders the ability of the skin to breathe.

Be prepared for behavioral changes in older llamas. Heavy-wooled males that have been shorn for the first time will actually "feel young again." Males may become more aggressive in breeding and defending their territory while sporting their new look.

Shearing may show you a llama that is an entirely different color than what you thought. Occasionally a white llama, when shorn, actually shows himself to be a dilute appaloosa (having a washed-out coloring of spots).

Grooming your llama before shearing is a matter of preference. Your decision depends on what you intend to do with the fleece. A general surface grooming is recommended because deep grooming may damage the delicate fiber. If you are breeding for fiber use, a more thorough grooming is needed. A matted fleece full of debris has little value. Shear a matted llama within an inch of the scalp. The fleece will grow out, and with weekly brushings or grooming, his appearance will be much improved. Even if you don't brush, the following year the fleece will be clean and healthier. Llamas that have not had regular grooming or shearing will need this start-over procedure.

Modified Blanket Clip

The most popular cut is the modified blanket clip. In this style, 2 or 3 inches of fiber are left on the "blanket" portion of the llama, leaving the legs, tail, and neck unclipped. A blanket clip only harvests the most usable portion of the fleece. The remaining areas are "feathered" or blended into the clipped blanket, creating an even, flowing look that still retains enough fleece to show what your llama is capable of producing. This is the cut you'll see on most breeding and show llamas.

The blanket clip takes practice, but don't be concerned if you make a mistake — the fleece will grow back. You may want to start with a gelding or one of your less valuable animals.

Cut path along topline

Step 1. *Start by standing to the side at the base of the llama's back just in front of the tail. Cut a shear's width from the tail to the base of the neck (withers) right along the spine. Be sure to hold the blades in a horizontal position. It is better to cut less than more here. You want to cut as straight a topline as possible. You can always go back and trim more off later.*

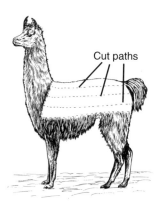

Cut paths

Step 2. *After having made the "path" cut, you will now see the fleece as two distinct halves. By shearing a path on alternating sides, you can mold the form of the animal through the cutting of the fleece. Use your nonshearing hand to gauge the distance of the cut from the body to keep from shearing too close and to keep the shearing an even length. Your free hand must continually feel for the body curves that you will follow with the shears. Remove the fleece after three passes or roll it back out of the way so you have a clear path. As you progress, the weight of the fleece will pull on the skin and cause too close a cut if you aren't careful.*

Increase length slightly at round of belly

Leave "skirt"

Step 3. *As you near the curve underneath the belly, increase the length slightly and leave a "skirt." This gives your llama protection from biting flies and insects on his belly when he is in the kush position.*

"Feather" clipped area into remaining area on legs to blend on long-wooled llamas

Step 4. *Clip short- or medium-wooled llamas right down to the top of the legs. (No blending is required.) A long- or heavy-wooled llama's fleece must be feathered at the neck, leg, belly, and tail to produce layering that blends evenly into the shorn blanket area.*

A properly and beautifully shorn llama is pleasing to the eye and shows off its natural athletic structure.

Step 5. *When the job is just about complete, stand back and look at your animal. You can now make second cuts to even out the coat. A Circuiteer is handy to blow out the shorn portion and will raise the fibers so you can trim evenly.*

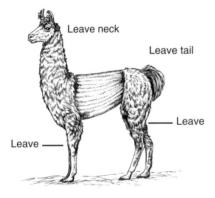

Leave neck

Leave tail

Leave

Leave

The barrel cut allows you to clean the back of the work or pack llama before you apply saddles and packs.

Barrel Cut

This is a popular cut for pack llamas and heavily fleeced llamas. The densest, longest fiber is around the barrel of the animal. This cut aids immensely in dispelling heat. Follow the basic "blanket" cut procedure, then cut the barrel area with electric shears. For a different look, clip or shear only in the barrel area.

Circulation Cut

This trim simply clears the natural heat dissipation areas of the llama: around the armpits behind the elbow joint and in front of the hips. These are smooth-haired areas that allow air flow and circulation. By increasing the air

flow here, your llama will find relief from heat and humidity. You can widen these areas by as much as 3 inches. You can cut this closely because the longer fleece will provide coverage. This is a good trim if you are going to show but don't wish to shear completely.

The circulation cut allows more air flow to aid in heat dissipation.

The Complete Cut

In this shearing, you'll remove all the fleece from the llama, leaving a uniform 2 to 3 inches. Follow the blanket cut procedure, except also shear the neck

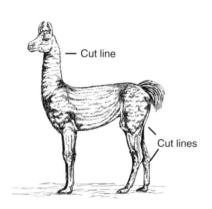

— Cut line

Cut lines

and legs. This cut is useful with a llama who has not been groomed or shorn in years and is covered with mats, felted fiber, and debris.

The complete cut is a must for "starting over." It is far less stressful to your llama than trying to save the fleece by trimming out mats and brushing thoroughly.

Advantages of Hand Shearing

- ◆ No source of electricity is needed
- ◆ Less noise to annoy the llama
- ◆ Blades are lightweight and easy to carry
- ◆ Easier to sharpen than power clipper blades; one needs only a hand stone
- ◆ Hand shears are less expensive

Fiber Preparation

Gale Birutta

These females have recently been shorn.

If you are breeding for fiber and fleece quality, your shearing skills will come in handy when the time comes to prepare the fleeces. You can prepare them at home, but if you have a number of fleeces, send them to a commercial mill. Depending on the cleanliness of the fleece, you can spin llama fiber directly from the shearings and wash it later. Many prefer to clean the fleece first; the choice is yours.

Hand Preparation

Some people consider hand preparation time-consuming and tedious work. But you may enjoy preparing the fleeces yourself, and the reward is a finer finished product. Be sure to clean the fiber by teasing it with your fingers until the debris falls out.

Hand carding opens the fibers, so you can remove any more debris and align the fibers for spinning. You'll achieve a nicely textured natural look when your fibers are hand picked and spun into the final product.

Rawna Gillette

Hand carders produce a fine, even yarn that is easy to work with.

Rawna Gillette

To process quantities of fleece, many people prefer to use a drum carder.

Commercial Processing

Many commercial mills now accept llama fiber for processing. You may have to deliver a certain amount — usually a minimum of 5 pounds — but it is more cost effective to process a large quantity. If you can't meet the minimum, try to join with other breeders to combine smaller amounts of fiber. Llama fiber cooperatives are springing up all over North America for just this reason.

Commercial drum carder.

Some mills have procedures to remove guard hair from the fleece if you are unable to do so. Mills will process your fleece as 100 percent llama or, at your request, they will mix other fibers with it. The processed fleece may be returned to you in batts; that is, in bundled layers.

Washing Fleece

Fleece is washed by soaking it in very hot water and plenty of soap or detergent — *without* rubbing or moving it about in the water. If it will be used for handspinning, choose only the best and cleanest parts. For felting, any quality of wool will suffice. Detergent cuts grease better than soap, and it also rinses out more easily. If you are washing a large quantity of fleece (20 pounds, for instance), you will need about 10 cups of detergent for a 20-gallon laundry tub filled to within 5 inches of the top with hot water (the hottest tap water you can get). Do not use your bathtub for this process, because the water will cool down too fast and the grease that the hot water lifts from the fibers will be redeposited on them.

Without taking a lot of time and effort, you want to

◆ Wash the maximum quantity of wool
◆ Get the wool very clean of grease and gumminess

Washing the wool. Pull the fleece apart, and shake out as much of the dirt and seeds as possible. This also keeps the shorn ends from matting during the washing process. Push as much wool into the hot water in the tub as you can get. Cover the tub to keep the heat in. Soak the wool for 1 to 2 hours.

Squeeze the washed wool as you take it out of the detergent water, and fill two mesh bags or old pillowcases with as much fleece as will fit into your washing machine. Place the bags into your washing machine, and run them through the spin cycle. Repeat with the rest of the washed wool.

If you are going to be washing a great deal of fleece, it would be good to have an old washing machine just for this purpose, rather than using the family machine.

Rinsing the wool. The wool is now ready for rinsing. If you expose the fleece to extreme changes in temperature, it will felt, so it is very important to have the rinse water at the same temperature as the slightly cooled-down wash water, but still hot enough to remove the dissolved gumminess. Discard the wash water from the tub and replace with clean, hot rinse water. This time, add just half of the washed wool to the rinse water. Squish the wool up and down several times, remove it, and put it in the mesh bags to spin out the rinse water in the washing machine. Repeat this procedure with the other half. If the wool is extremely dirty or greasy, you may have to give it two rinses.

Drying the wool. Fluff the wool, and put it out on wire racks to dry. Rust-free chicken wire makes a good drying rack When it is dry, put it in a cloth bag or a cardboard container, seal well, and store until use. Old pillowcases without holes allow ventilation but keep out moths.

The Future of Alpacas

There is only a small population of alpacas in the United States — close to 10,000; their numbers will have to increase substantially before a commercial market for alpaca fiber can be established. Just a few custom mills process alpaca fleece; some are located in Pennsylvania and Michigan, and are listed in Appendix B. It is not yet economically good business to try to make a living from alpaca fleece — it's just too costly to process and too expensive for a large market. However, alpaca breeders are upgrading their herds and laying a good foundation for a commercial industry to develop.

Marketing Your Fiber

With the growth of cottage industries and the high quality of llama fiber, marketing possibilities are almost endless. Most llama shows now have either fleece-on-the-foot competitions or fiber and yarn judging. They also hold competitive fashion events to highlight the versatility of this wonderful fiber.

Labeling

Before starting your marketing plan, be familiar with the Textile Fiber Products Identifications Act of 1960 and the Wool Products Labeling Act of 1939. Both mandate that the percentage of each type of fiber, instructions for care, place of origin, "manufacturer," and price be affixed by label to any item for sale. This is also required for raw fleece. If you intend to mix your llama fiber with other fibers, either at home or by a mill, you must know the percentage of the fiber mix so you can label properly.

Selling to Spinners

Fleece must be of the finest quality to be marketed to hand spinners. Cleaned and carded fleece sells for $2.50 to $4 per ounce — that's $40 to $64 per pound. Black, gray, reds, and fawns command a higher price. You can offer discounts when someone buys larger quantities.

If you are not a spinner, perhaps you know one who will barter spinning services for raw fleece. You'll be one step closer to your own finished products.

Provide a custom *scouring* (washing) service for your spinner customers. Careful scouring of a fleece for hand spinning could double its value, and this process does not require expensive equipment.

Yarn Shops

There is probably a craft or yarn shop nearby. Stop in and introduce yourself. The shop may be interested in displaying your fiber or yarn on consignment. The yarn shop will need to make a profit as well, so you will be selling to it at wholesale prices. Wholesale prices range from $3 to $5 per ounce; the retail price ranges from $7 to $14 per ounce, depending on the color and quality of the wool.

Package fleece in clear bags to show off the assortment of natural colors. Combining more than one color in each package is also attractive. Amounts of 2 to 5 ounces are generally more salable.

Spinning, Weaving, and Crafters Guilds

Guild meetings offer you a captive audience. Crafters will pay top dollar for specialty fibers of excellent quality and unusual colors. They are knowledgeable and talented consumers, so they recognize and demand the very best.

Market to guilds during the spring shearing season and in the fall. For the fall season, dye wools in earth tones. Create special blends for the holidays. Join the guild, attend its regular meetings, and advertise in its newsletter.

Public Events

By holding "open barns," shearing demonstrations, and clinics, you will introduce new people to the world of llama fiber. Compete in fleece-on-the-foot competitions at llama and alpaca shows — you'll be in the public eye, which will benefit your farm, particularly if you win the class. Enter fleece and yarn at area fairs and farm shows to promote your fiber and your business.

Whenever possible, have llamas on display, along with clean and carded raw fleece, spun yarn, and a variety of finished products. If you are not a spinner, make arrangements to have a spinner present and working. By exhibiting an exceptionally clean llama, the spinner will be able to spin directly from the animal's fleece without processing it first. Display a full pound of clean and carded wool to demonstrate how much is really in a pound of llama's wool.

FOURTEEN

ORGANIC FERTILIZER

Llamas produce a high-nitrogen, naturally rich manure that can be applied directly and immediately to plants without the danger of burning. Llama manure is easy to process from the raw stage to a highly sought after organic fertilizer for the retail market. As a bonus, your pastures will be cleaner when you collect your "products" for processing.

This chapter will discuss all phases of fertilizer production. But first let's review the basics of plant nutrition, so you will undertand how and why llama manure forms the foundation of a high-quality fertilizer that plants will appreciate.

Plant Nutrition

Certain elements provide plants with an adequate supply of nutrients. Plants need carbon, hydrogen, oxygen, nitrogen, phosphorus, potash, magnesium, calcium, copper, boron, manganese, molybdenum, iron, sulfur, and zinc. With the exception of carbon, hydrogen, and oxygen, plants must be able to absorb minerals from the soil through the roots. We are going to focus on nitrogen, phos-

Kevin Kennefick

Odorless, llama manure is naturally high in nitrogen but will not burn plants.

261

phorus, and potash (potassium). These nutrients are essential for plant growth.

Nitrogen

Plants require nitrogen for the production of leaves and stems. However, too much nitrogen could result in an overabundance of leaves at the expense of fruit or flowers. Stems will weaken and rot, and the plant's resistance to disease will be lowered.

Soils are at their lowest nitrogen levels during early spring because heavy spring rains tend to deplete the soil of its nitrogen content.

Phosphorus

Phosphorus is vital for the development of flowers, seeds, and fruits — stunted growth and sterile seeds can result from a deficiency. Stable root development, sturdy cell walls, and early maturity all depend upon adequate levels of phosphorus. Phosphorus may also boost the vitamin content of plants. Unlike nitrogen, phosphorous does not leach out through the soil.

Potash (Potassium)

Potash promotes the all-around vigor of the plant and increases its resistance to disease. It also generates sturdy root growth, decreases the need for water, promotes fruit and flower color, and assists plants in utilizing nitrogen.

Fertilizer

Lawns, gardens, houseplants, flowers, shrubs, and trees all need a complete and balanced fertilizer. Most organic fertilizers contain an average of 3 percent nitrogen, 2 percent phosphorus, and 1 percent potash. Depending upon the type of growth you wish to promote, these percentages or multiples of them are adequate.

In the analysis of a commercial fertilizer, the first number always represents the nitrogen content (N); the second, phosphorus (P); and the third, potassium (K). Most manufacturers of "complete" fertilizers offer two forms: the quickly absorbed inorganic form, and the more slowly absorbed organic form. In the organic form, nitrogen is available to the plant over a longer period of time, which means you don't have to fertilize as frequently.

Fertilizer Components

A basic understanding of fertilizer components and how they work will help you develop your own product.

Animal Manures

Animal manures were used long before commercial fertilizers came on the market. In addition to increasing soil fertility, animal manures will:

- improve the physical structure of the soil
- increase the organic content of the soil
- increase the bacterial activity of the soil

The average percentages of nutrients for various types of manures are as follows:

	Nitrogen (N)	Phosphorus (P)	Potash (K)
Chicken Manure	1.0	0.8	0.4
Horse Manure	0.7	0.25	0.55
Sheep Manure	0.95	0.35	1.0
Cow Manure	0.6	0.15	0.45
Pig Manure	0.5	0.35	0.4

As low as these percentages of vital nutrients are, only about half of the nitrogen and phosphorus, and approximately one-sixth of the potash, are actually available for use by plants. Thus, the actual value these manures add to the soil is minimal. By utilizing llama manure, you can develop a viable fertilizer without usually having to add supplements.

How you handle animal manures is extremely important. To preserve fresh nutrients, store manure under cover or keep it in containers to retain its natural moisture. If the manure dries out, it will be useless to soil and plants.

Fresh manure should not be used in direct contact with the roots of plants; it can burn tender roots. Partially rotted manure works well: Higher nutrient values are available in this form, and any burning effect is at a minimum.

Llama manure is a happy exception: You can apply it directly to plants. If it's going to come in contact with roots, however, mix four parts llama manure to one part soil.

Dehydrated Manures

There are various forms of dehydrated manures (the most common is cow manure); however, they are largely useless as fertilizers because during the drying process, most of the valuable nutrients are lost. Humus and commercially based fertilizers are a better bet unless the dehydrated manure contains additives.

Liquid Manure

Originally, gardeners would immerse a bushel of manure in a barrel of water and would use the resulting "tea." In modern applications, nitrate of soda, lime, and ammonium sulfate or a complete fertilizer is dissolved in water and applied to the plant. A complete fertilizer is one that contains all of the main nutrients — nitrogen, phosphorus, and potash (N–P–K).

Organic Additives

Additives will increase the amount of nutrients to their proper level:

Cottonseed meal: Cottonseed meal usually contains as much as 7% nitrogen and delivers its nutrients over a long period of time.

Wood ashes: Wood ashes contain approximately 1.5% phosphorus. They are also an excellent source of potash, containing 97% of this nutrient.

Natural rock fertilizers: Granite dust, greensand, and rock phosphate contain as much as 30% phosphorus. Granite dust also contains approximately 6 to 7% potash.

Grass clippings: These are fairly rich in nitrogen content, and can be added to a composting process.

Hulls and shells: Shells and hulls of cocoa beans, rice, buckwheat, and cottonseed will increase nutrient value. These contain 1% nitrogen, 1.5% phosphorous, and approximately 2.5% potash, and are available from stores, nurseries, and mail-order houses.

Leaf mold: Leaf mold consists of damp, decaying leaves. Although leaf molds differ, nitrogen content can be as high as 5%. To make leaf mold, place shredded leaves in a container and keep the pile damp. In the spring, you'll have a fine supply of leaf mold.

Sawdust: Sawdust is very low in nitrogen, containing only about .01%. Sawdust is useful, however, if the fertilizer you are producing is too wet (probably from urine) and needs to be dried a bit — it will absorb excess moisture.

Seaweed and kelp: Seaweed and kelp contain approximately 5% potash and small quantities of other trace elements. Dehydrated forms are available commercially if you cannot harvest it directly. Keep in mind that the salt content in any commerically prepared supplements has been removed during processing. But if harvesting fresh kelp and seaweed yourself, you first must thoroughly rinse the harvested plants to avoid applying excess harmful salt to your garden.

If you choose to add nutrients such as natural rock fertilizers or cotton-seed meal to your llama manure fertilizer, remember that such additives will add to your costs.

Processing Methods

Which method of processing is best? The answer depends upon the amount of time and money you are willing to invest in your fertilizer venture. Try each of the following procedures to determine which is best for you.

How many llamas do you need to do the job? For a composted product, 8 to 10 llamas produce enough manure to give you 200 2-pound bags of fertilizer per month, or 400 pounds. Don't be concerned about not having enough manure; other owners and breeders will gladly supply you with all you can handle.

Drying

Most llama owners go with this method, as drying makes a much more consistent and attractive product to handle. Before drying, mulch the manure; without mulch, whole pellets will turn into rock-hard marbles. Any odor is minimized in the drying process. Drying decreases nitrogen content, though, so you'll need additives to enrich your product.

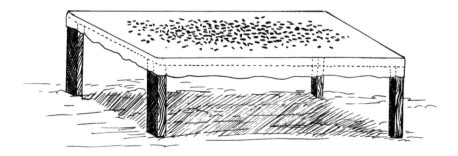

Solar drying works well if no rain is predicted for several days and there is consistent sun.

Drying can be accomplished several ways, one of which involves ovens. In this case, you will need a commercial-style oven to use for this purpose only. Electric ovens are best. Place the manure about 1¼-inches thick on old cookie sheets and bake at 325°F for about 15 minutes. Keep the oven in your barn or outside to avoid living with the odor that baking produces. You can use a microwave, but it's a slow process because these machines are so small.

You may want to try solar drying. With this method you can dry substantial quantities at once. Here, too, the pellets must be mulched. Lay a canvas tarp (not plastic) over wood ties so the tarp is slightly off the ground. You can even lay the tarp over a picnic table. Spread about 1¼ inches of the mulched manure on the tarp and cover it with another canvas tarp. Drying time depends on the weather, but should take from 2 days to a week. If it is very sunny and the air is still, uncover the pile. This will speed drying time considerably.

Drying Equipment

You'll need:

- an electric oven. You can get a good used residential-sized oven for approximately $200.
- cookie sheets or pizza trays: aluminum or stainless steel; stay away from nonstick sheets. Approximately $6–$8 a piece; check with pizza shops for used ones.
- canvas tarps (two): For solar drying, 6 x 8 feet or 10 x 12 feet will do. The larger the tarp, the more manure you can dry. Tarps can cost from $15 to $60 each, depending on size and quality.

The manure is dry when it feels like stale coffee grounds between your fingers. Have the manure analyzed at this time to determine its nutrient content. If you are marketing your product as a true organic fertilizer, then this is an important step. But if you plan to use it yourself or give it to friends, you can skip the analysis.

The advantages of drying are:

- The manure's texture is consistent.
- The finished product is pleasing to the eye.
- The product has little odor.
- Drying requires only a small investment for equipment (if you are solar drying).

The disadvantages are:

♦ The product is low in nitrogen.
♦ The purchase of a stove is expensive.
♦ The procedure is time consuming.
♦ It results in less weight-to-volume ratio.
♦ The product will need additives.

Composting

Another processing method is composting. The key to successful composting is to keep it sealed or covered to retain nutrients.

Using simple plastic trash containers, you may compost the pellets first and then mulch them, or mulch them first and then compost. With either method, you must have the product tested for nutrient value every six weeks. The analysis may be quite different if you mulch first and then compost, or vice versa (temperature and moisture levels will affect the analysis). Decide on your method and then be *consistent*.

Composting Equipment

Composting containers: You can use plastic trash cans or sophisticated composters. Specially designed composters will be most efficient, but the drum-style composter, for example, can run as high as $1500.

Plastic trash cans, on the other hand, can be purchased from your local hardware or feed store for $8 to $15. Whichever container you choose, also add a good composting book to your library.

The advantages of composting are:

♦ It produces a high nutrient content.
♦ It requires little effort.
♦ It creates a large quantity.

The disadvantages are:

♦ Sophisticated composting equipment or tumblers are expensive.
♦ Cold climates may dictate the time of year for processing.

"Fresh" Manure

Manure can be used fresh — it has been used this way for centuries. Horse, sheep, cow, and chicken manures are considered hot. Hot manures are those that are exceptionally high in nitrogen and therefore have a tendency to burn the plants. These hot manures are normally more moist than other manures. Generally, the wetter the manure, the higher the nitrogen content.

All manures can be used in organic fertilizers. Llama manure can be used as a "time-release" product because of its naturally pelletized form. If you are marketing on a commercial scale, you must check your state regulations. For instance, Connecticut requires that manures sold in bulk by the distributor or farm be picked up by the user. This even holds true, for instance, if someone wants to buy a pickup truck–load of manure for his garden.

The advantages of raw or fresh manure are:

+ No processing is necessary.
+ The manure will go farther as there is no processing weight loss.

The disadvantages are:

+ The nutrient content is inconsistent.
+ The raw urine in it causes high odor.
+ Its moisture promotes fungal growth.

Liquid Manure

Liquid manure is an excellent source of nutrients. Brew a "tea" by placing mulched llama pellets in a container with water. After the contents have steeped a while, strain out the pellets. Water houseplants and garden vegetables with this tea. They'll appreciate the nutrient drink.

Liquid Manure Equipment

You can use simple plastic gallon milk jugs or any other disposable container with virtually no financial investment. Keep in mind that these containers should be white plastic or frosted in color to alleviate the reduction of fertilizer nutrient values caused by too much exposure to light.

You will need to "collect" the pellets and keep them as pure of hay and bedding as possible. This is not critical, but makes it easier to process.

Try installing a concrete pad for your llamas to be potty trained on. A pad will make collection of pellets much easier and cut your processing time by eliminating hay bedding. (This may be easier said than done.) If you are lucky enough to be introducing llamas to a virgin pasture, you can install the pad and place some pellets on it. This will encourage the llamas to consider the pad their dung area. But if you are working with an existing pasture and your llamas already have a dung area, remove the present dung pile and install the pad where the animals were making their deposits. A 4- x 4-foot pad made from plastic or wood will suffice.

What about All These Pellets?

I heard of someone in the western part of the country who made fertilizer from llama manure. I tracked him down and sent for a sample of his product. When the package arrived, I opened it up to take a look. The product had been dehydrated, was of a pleasant texture, and had no odor. But what about its nutrient content? My research shows that dehydrated manure is of little value because almost all the nitrogen has been lost in the drying.

I made my mind up that if I was going to "manufacture" a llama manure–based fertilizer, it would not be dried and it would keep its naturally high nutrient content. I experimented with batch after batch and test after test, and finally came up with a composted version with a ratio of 5:3:3 (N-P-K). Because I applied for the prestigious Vermont Seal of Quality, this could only be marketed as a guaranteed analysis product. To give myself a buffer zone, I guaranteed the product at a lower analysis of 3:2:1 (N-P-K). The product took off and I began a good-sized mail-order business. Even specialty stores in New England stocked my fertilizer.

Testing

First try your local extension service or the Department of Agriculture. If there are regulations in your state regarding the sale of manure or fertilizer, you will need to find an independent lab to run your analysis. A local agricultural technical college is likely to have facilities for performing such an analysis.

It is important that you continue to test your product through all stages of development and processing to keep tabs on its nutrient content. The nutrient rating may not be consistent. Be prepared to spend $15 to $50 per analysis.

State and Federal Regulations and Licensing

Consult with your state's department of agriculture about any laws and regulations regarding the processing and sale of manure and fertilizers. Each department of agriculture is a state branch of the U.S. Department of Agriculture (USDA), so by complying with state regulations you are also complying with those of the USDA. Contact the Plant Industry section of the agriculture department and request a copy of the rules and regulations for marketing, processing, selling, and licensing commercial fertilizers.

By marketing your fertilizer as a "specialty fertilizer," you may be exempt from many of the regulations. Check to see if that's the case in your state. If you plan to market your fertilizer in states other than your own, check their requirements for licensing. Licensing fees vary from state to state.

Abide by your state's laws and regulations. Do not hesitate to ask questions. Take the time to understand the requirements — you want to avoid a fine or revocation of your license for noncompliance.

Packaging

If packaging is regulated by your department of agriculture, there will be standards to which you must adhere. Some states will need to approve your prototype packaging and label. Check with your state's department of agriculture. Some control the way packaging is to be sealed and the colors of labels.

Registering a trademark also involves many regulations. The process of obtaining registration can be time-consuming and annoying, but pays off in the protection it affords. The type of packaging depends on your market. If you would like to target the upscale buyer, consider a 2-pound gourmet coffee–type bag with a plastic liner. A colorful and eye-catching package will stand out (as usual, incorporate your farm colors). It is worthwhile to make the extra investment here.

The type of processing you have chosen will also influence your packaging. Whether your fertilizer is very moist or very dry, I would recommend

This fertilizer product manufactured in Vermont is endorsed by the state's department of agriculture and carries two registered trademarks.

using a plastic liner inside the package to prevent problems and to make a tidier seal.

Packaging must be designed to withstand handling. Cardboard boxes, cans, and see-through plastic bags are all possibilities. Keep in mind that you are marketing an organic product — don't throw your organic image out the window by marketing your product in something that is not environment-friendly.

Logos

Employing a professional to design a logo or trademark can be costly. Contact a local high school or art school to see if any students would be interested in your project. Students will welcome the opportunity to show their talent, and they'll have an addition to their portfolio. Market studies indicate that logos in a circle or oval are the most pleasing to the eye and the most memorable. Keep the logo simple, uncluttered, and recognizable.

Marketing, Promotion, and Advertising

The success of your product lies in its marketing and promotion. Always refer to your product as *fertilizer*. It's no longer raw manure, and some people are squeamish about the word "manure."

Marketing Research

Gather information on organic fertilizers that are sold in your area. Stick to the "specialty market," because that is where your product will be sold. (Don't compare your product with the 50-pound bag of Agway fertilizer that sells for $9.95. This is not your market, nor is it your competition.)

Send for free garden and seed publications to see what they are advertising in the way of fertilizers and what prices they are charging.

You must know your competition. Make note of the contents of competitors' products, pricing, where they advertise, where their product is sold, their guarantee, packaging, *everything*. Approach the development and goals of your product with the upscale shopper in mind; you won't have as many customers, but they are willing to pay more for a fine product. Attract organic gardeners wth your advertising — they will appreciate the value of your fertilizer.

Decide whether you want to process 100 percent (pure) llama manure or if you want to add various supplements to enhance the nutrient quality. If you want to sell *all organic*, remember to utilize only organic supplements. Other supplements are chemically based, such as triple or super phosphate.

Target Your Market

Organic gardeners are your market. Your advertising and packaging should reflect their needs. Think of your product as a consumable, something that will be used and replenished. Look at your pricing structure and decide if you want to cater to the houseplant lover or outside gardener. Cover both markets by offering your product in both small and larger quantities. A 2-pound bag is perfect for houseplants; a 10-pound bag will take care of the small backyard garden.

Organic gardeners are concerned about content: They will scrutinize the labeling. It won't matter if it is llama, horse, cow, bat, or moose fertilizer — the organic gardener's number one priority is *organic* matter.

Marketing and Promotion

Market your fertilizer in:

- organic gardening publications
- garden clubs
- mail-order gardening-supply companies
- upscale organic garden centers
- country stores, general stores, and specialty shops

Write to these places (the cost of telephone calls can mount quickly). By using letterhead stationery, you present a professional appearance to the potential retailer/purchaser. Avoid photocopies. Each letter should be addressed to an individual retailer and contain the manager or owner's name. Follow up with a phone call. Be sure to ask for an appointment to show your product and its advantages.

Write press releases. The business reporter of your newspaper is always looking for a good story, and a new local product is newsworthy. As a bonus, he'll be mentioning your farm and the other enterprises in which it is involved. This is useful and interesting information for the public, and just happens to be great publicity for you.

T-shirts are another way to advertise your product. Give a T-shirt to every retailer who sells a certain amount of your product.

Check with your state's agriculture department. The folks there may have some marketing and promotional ideas for local producers; there may even be a fulltime marketing specialist who can assist you.

Consignment Sales

Many retailers will be happy to accommodate your fertilizer on a consignment basis. Make an appointment with a retailer to discuss selling on consignment. Dress professionally to meet her. A neat appearance gives the retailer confidence in you and your product.

You supply the retailer with your product at no charge. The retailer pays you when the product is sold. Be sure to include a consignment agreement for retailers to sign; you do need some protection. A 20 to 40 percent commission to retailers is fairly standard. As the product starts to sell and demand starts to increase, lower the commission percentage. Make it clear to the retailer (in writing) what price the product is to be sold for. You do not want retailers undercutting each other.

Printed Material

Brochures are a great marketing tool. If you have confidence in your writing ability, draft something simple on your own. If you have a computer, you can design a brochure with any of the desktop publishing programs, or simply compose a one-page flyer on a word processor. Many word-processing programs now incorporate an easy-to-use desktop publishing component. You can always find someone to help.

Target the direct retail customer as well as the wholesale buyer. Incorporate the wholesale and retail markets into one brochure, making an insert to accommodate each one individually. For wholesalers, add "Quantity discounts available to retailers." For retailers, add "Get a 10% discount with your first order." You can produce a professional-looking brochure with all the information on an 8½ by 11 page, printed on both sides in three-column format, and folded in three. In this way, the printer only has to run it through the printer twice (if one color). About 500 copies of camera-ready material, printed on both sides (unfolded), can cost as little as $60. Check with your local print shop for suggestions. It may also help you with layout and design.

Paid Advertising

Advertise in the classifieds of pricier mail-order catalogs. This is where the serious shoppers will be looking. If you want to invest in a block ad, get together with several of your retailers and share the cost of an ad announcing that your product is now available through those retailers. The sales you generate in this way will more than repay the cost of advertising.

Radio/Television

A local radio station may benefit your product. Radio talk shows have many listeners; contact the station to see if you fit in with a topic they have on tap. Check with your local Public Radio office as well. Like newspapers, radio stations are always looking for interesting people to interview. If you're not shy about speaking to the public, you'll reach hundreds of people — who knows, you may become a media star. The same applies to local public access television as well.

Find out from the nearest chamber of commerce what trade or garden shows are held in your area. Apply to display at these shows.

Mail Order

The initial investment of ads in mail-order magazines is high, but a well-planned ad will bring remarkable results. Contact a professional marketing firm for assistance. They'll know the key words that will bring attention to an ad.

With mail order, be prepared to have enough product to meet demand. Keep track of where your mail orders are coming from. This will enable you to know what ads in which magazines or publications are bringing the best results. Most mail-order customers are impulse buyers, so contact your bank about accepting MasterCard and Visa payments.

Provided you have the drive, ambition, energy, and self-confidence to market your fertilizer, you will be rewarded with the outcome. You will make new friends in the agriculture world, and the additional promotion will benefit your farm and animals.

FIFTEEN

LIVESTOCK GUARDIANS

Llamas make excellent livestock guardians. This fact was discovered by accident when sheep producers who wished to diversify their fiber business housed llamas with their sheep. Sheep producers noticed that they were losing fewer lambs — thanks to llamas.

Gale Birutta

To diversify their fiber production, sheep breeders occasionally purchase a llama for its fiber and pasture the animal with their sheep.

Research Studies

Dr. William H. Franklin of Iowa State University was the first person to conduct formal research into the use of llamas as guard animals. In the early 1990s, Dr. Franklin studied how llamas guard and why, how they compare with other guardian species such as dogs, the longevity of guardian llamas, and the cost savings of this approach to predator control. In 1995, Utah State University established the Predator Ecology Project to research ways to help producers cut their losses to predators while not harming the predator species. The efficacy of guard animals, electronic devices, and various types of fencing are all under review.

The Guardian Llama Placement Program in Vermont utilizes private funds to continue the research begun by Iowa State University. The Vermont program studies llamas working with guard dogs, llama guards working in pairs, and breeding specifically for guardian llamas.

Why Llamas Guard

Llamas have a natural dislike of canines (domestic and wild dogs, coyotes, and foxes). For thousands of years, members of the canine family preyed on young llamas in South America. (Llamas thus have evolved to be very protective of their young.)

When a llama is removed from his llama "family group," and following a proper introduction to and bonding with the new livestock species, the llama takes over as protector and leader in his new family group (sheep, goats, or other livestock) and establishes exclusive territory for his charges. Research is under way to study the question of why a llama takes on this responsibility.

How Llamas Guard

Highly intelligent, llamas can distinguish between possible and imminent threats. Unlike guard dogs, which may attempt to kill anything unusual in their territory (including domestic pets), llamas become aggressive only if a predator (nonhuman) advances toward their new family group. Some llamas will even allow families of foxes to live in pastured areas with ewes and lambs so long as they pose no threat to the flock. Coyotes, too, may be safe in a guarded sheep pasture, but only when they pose no threat to the flock.

A confrontation between a llama and a predator will usually result in the predator's retreat. When the predator approaches the llama or flock, the guardian llama positions himself between the flock and the predator. He'll show aggression by chasing, biting, and kicking. Guardian llamas can stomp to death a coyote or a dog that won't retreat. A llama may also herd the flock to a protected area, such as a building. Llamas have even challenged moose and herded sheep into barns when planes or helicopters have approached.

Gale Birutta

Llamas are alert and will sound a shrill call of alarm when they spot trouble.

Guardian llamas protect their families in two ways. The most common is by keeping the flock in a tight group and staying nearby. Some "patrol" the fence for any incoming predators or find the highest ground from which to observe the flock. As the flock moves, some llamas take on the role of leader, bringing sheep to and from pasturing areas and housing; others take up the rear and follow the flock, making sure there are no stragglers.

Issues with Guardian Llamas

Most of the problems associated with the use of guardian llamas arise from a lack of understanding of llamas' psychological traits. Little is known about this, as most of the study of llamas with other species only began in 1995.

Young Llamas as Livestock Guardians

While young llamas (between the ages of 6 and 18 months) may guard sheep, they are not effective in an extreme predatory situation. It seems that although the young llamas immediately bond with the flock, they also run away from the predator with the flock. When llamas are removed too soon from their families, they tend to become like the livestock they have in fact been assigned to protect.

Young llamas need to learn the ways of their own species and mature into young adults in order to develop the territorial instinct that will make them valued guardians. Between 18 months and 2 years of age, the llama is taught by his more aggressive peers to be territorial. Now he is ready to guard other animals.

Intact Male Llamas as Livestock Guardians

Intact male llamas should never be used as guardians. They may attempt to breed a ewe, which can kill her. One intact male llama killed 100 ewes before anyone realized why they were dying. Gelded males are the best choice, provided they were not used for breeding before being gelded. Females are excellent guardians, but they are rarely used because of their value as breeding stock.

The Ideal Guardian Llama

Which llama will make an ideal guardian llama? We've already established certain criteria. The llama *must* be gelded, preferably at 18 months to 2 years of age (before his hormones become active). The llama should have a "loner" type of personality. His territorial instinct must be well developed. A llama that is accustomed to other types of livestock is preferred, as well.

It is recommended that a llama be housed previously with the type of livestock he will be assigned to protect.

It's always possible that a llama that appeared to be an excellent candidate for guardianship nevertheless fails at his duties. The llama may refuse to bond, jump over the fencing, or even injure the livestock he should be guarding. On the other hand, some llamas that are at the bottom of the pecking order will make excellent guards when removed from other llamas; they may immediately develop a territorial instinct and a take-charge attitude.

Even though most llama breeders will exchange a llama if a problem arises, promises such as "I'll replace the llama in 30 days if he doesn't work out" are of no value if losses continue while the problem llama is still there. Try to acquire an experienced, working guardian llama.

The Guardian Llama Placement Program

This program began in Vermont in January 1995 to screen llamas as potential guardians. The program acquires llamas from breeders and screens each first by housing the animal with sheep, goats, bovines, poultry, deer, and other livestock species. The llama is closely monitored for a period of 30 days at the research facility. At the end of a month, the llama is then placed into "field" training. He is shipped to a volunteering alternate livestock facility, where he again will be monitored for 30 days. This tests the llama for reactions to changes in environment. After meeting an extensive list of criteria, the llama is available to producers of the livestock that llama works best with.

Criteria Checklist

♦ Is the llama staying with the flock or herd, or does he wander away?
♦ Does the llama alertly and actively investigate anything out of the ordinary?
♦ Is the llama respectful of fencing?
♦ Does the llama routinely bond with lambs, kids, or younger livestock?
♦ Does the llama overly interfere with birthing?
♦ Does the llama allow the flock to graze sufficiently without constant "rounding up"?
♦ Does the llama pace the fenceline without regard to where the flock is?
♦ Is the llama quick to sound an alarm call when danger nears?
♦ Is the llama aggressive toward people or pets?

The livestock producer receives a working guardian llama who does his job from the first day, because the llama is already comfortable with that species and is a proven guardian.

The program is also experimenting with using llamas in pairs. Pairs are advantageous when a major predation problem arises. Even the best guard llama cannot be everywhere at once. By working in pairs, one llama will keep the flock together and be on the lookout, while the other can patrol the remaining area or the fenceline.

The Guardian Llama Placement Program works in conjunction with various agricultural departments, extension services, and animal damage control services nationwide. (The national headquarters continues to operate in Vermont, handling placements throughout New England.) Many states now also have guardian llama programs of their own. Currently, the New York chapter and Texas chapters are in full operation. Virginia, New Mexico, and California chapters are being negotiated. And it is the program's plan to eventually operate a chapter in every state.

Introduction to Other Livestock

Llamas have distinct personalities, so how a llama reacts to a new animal will depend on the individual animal. And a llama who is accustomed to sheep probably won't be a good guard with goats. Sheep are usually fairly docile and take well to herding. Goats, on the other hand, have minds of their own; they don't like to be herded. A llama must protect goats by deterring predators, not by shepherding them away.

First introductions may go in a number of different directions. For the most part, the llama is curious and just wants to become part of his new group. He may appear aggressive, but in fact he is just displaying his eagerness to be accepted as quickly as possible.

When a llama meets sheep, it may even appear that he is chasing them. What is really happening is that the sheep get frightened and run away. As the curious llama tries to get closer, they keep moving away, so it looks like he is chasing them. Remove a few sheep or other species from the rest of the flock or herd and put them in a smaller holding

Noah alertly guards Pygmy goats in Massachusetts.

In this introduction, the new owner holds the llama while the sheep are scattered. Eventually some of them will approach.

facility. Then introduce the llama in the confined area. Monitor the animals closely to prevent panic. Certain breeds of sheep are prone to more overexcitement, such as Montadales; others, such as Columbias, are more secure and are not usually bothered by llamas. Most breeds of sheep will accept the llama within several hours or days. Because of the behavioral differences among breeds of sheep, some llamas may be capable of guarding one breed of sheep but not another.

When llamas will be used to guard horses, introduction is best accomplished with the animals placed in adjoining pastures. Horses could seriously injure themselves or the llama while kicking or running. A successful introduction between a llama and horses may take up to a month.

Llamas as Guardians Compared to Other Species

Llamas make better livestock guardians than do dogs and donkeys. For one thing, llamas are tall (up to 6 feet) and therefore have excellent visibility above the flock or herd.

Llamas also require little bonding time; dogs may require 3 years. Llamas do not usually injure or harass their charges, humans, or pets. Llamas are safe around children. While dogs have been known to roam away from flocks, llamas will not leave their new family group. In addition, llamas will graze and browse with the livestock species, and don't require special feeding. A guard llama will do his job effectively for up to 15 years; dogs only do so for an average of 2 years. According to Dr. Franklin's research at Iowa State University, most guard dogs become ill and die prematurely or become seriously injured. This is because dogs are more susceptible to diseases from other canines due to fights or bites. Guard dogs pick fights with just about any animal, predatory or not.

Donkeys make good guardians, but many of their owners have switched to llamas. A donkey tends to lose interest in the flock within 4 years. Donkeys are playful, but their roughhousing may seriously injure or even kill lambs unintentionally. Llamas are gentle with young animals and in fact will actually intervene to speed the bonding between the lamb and the ewe. Llamas will push young lambs toward the ewe if the newborn seems confused. Donkeys must be removed during lambing to prevent trampling; llamas are harmless and can remain. And llamas do not chew on fencing, barns, or gates.

Working Ability with Guard or Herding Dogs

It was thought that llamas couldn't be used in conjunction with guard dogs or herding breeds of dog. However, recent research by the Guardian Llama Placement Program shows that certain llamas willingly accept guard and herding dogs and will still perform their duties. These llamas also remain effective guards against outside canine species. This program has successfully placed guardian llamas with various breeds of herding dogs, such as Border collies and Australian shepherds, and guard dogs, such as Maremmas and Great Pyrennes.

Nonlethal and Humane Approach to Predator Control

With the growing emphasis on the environment and nonlethal predator control, llamas are quickly becoming the guard animal of choice. Poisoning, shooting, trapping, and other lethal means of predator control can be curtailed or even eliminated when guardian llamas are on the job.

Coyotes and Other Predators

Coyotes are the most opportunistic predator in the United States. In fact, they prefer the easy way and are quickly deterred if the task of a kill is too dangerous. The most recent studies by the Animal Damage Control and Fish and Wildlife Departments show that changing the feeding habits of the coyote will solve the problem of predation without killing them. By effectively removing the coyotes' food source, one can encourage the coyote to feed on small wildlife rather than livestock.

When llamas are guarding a flock or a herd, a coyote will look elsewhere for a meal. Unfortunately, this may drive the coyote down to the neighbor's sheep farm. (This is why a livestock producer may not incur losses for many years, and then be hit with many kills over a short period of time. If you don't have a llama protecting your herd, the addition of guard dogs or electronic devices will also encourage coyotes to move on.)

Llamas have even successfully deterred black bears in the East. However, llamas themselves are vulnerable to predation by wolves. Guard llamas in the Pacific Northwest and Canada have been taken by packs of wolves. Even full-grown llamas in breeding facilities are not safe from wolves in the Northwest. In South America, the llama's only natural predator is the puma (also known as the mountain lion or cougar; we know it as the catamount in the Northeast). If your livestock is threatened by wolves or cougars, don't use a llama as protection; he will take flight. Immediately contact your local game warden or animal damage control center. Presently, the re-introduction of wolves into some western areas, such as Yellowstone, is creating friction between livestock owners and animal rights activists. Llamas are no match for mountain lions or wolves.

Number 635 Was Not a Lucky Number

A 230-pound registered Columbia ewe, tagged no. 635, fell prey to a black bear at our farm while John and I were in Pennsylvania hoping to locate additional potential guardian llamas for the Guardian Program's screening process. The unguarded ewe was taken several days after the llama guarding the flock was removed and placed as a guardian at another producer's request.

David Coughlin, a friend and neighbor tending to the farm's livestock in our absence, found one of the sheep missing early one evening. Other llamas on the farm were noticeably upset: pacing the fenceline, "alarm calling," and not eating. Dave found several of the fencing's fiberglass interim poles pulled up. The polytape utilized for the sheeps' rotational grazing was also down and pushed over to the side of the perimeter fence. Bits of wool were scattered about; they led down the pasture to the river's edge bordering the forest. At the end of the fenced pasture a pool of blood was found, the site of the final kill. Dave, an avid hunter, followed the trail back through the electric fence, over the river embankment and into the river, back up the other side to where a portion of the sheep's carcass remained; the hindquarters of the sheep had been eaten. After realizing that he was at the site of a recent bear kill, Dave left to return home to get his video camera and a weapon. Upon returning, a video tape was made and the remaining sheep were caught and put back inside the polyfence. Dave returned the following day and found that the carcass had been moved once again, further up the hill with only a leg and fleece remaining.

With no apparent signs of a struggle and three large droppings of scat (approximately 8 inches across and 1½ inches high), the overall feeling was that a very large black bear had taken the ewe. She has been dragged approximately ½ mile from the pasture to the final resting site.

The game warden was notified and a bear kill was confirmed; the first reported livestock kill by a black bear in Vermont for some time. The estimated size of the bear was between 500 and 600 pounds. Our farm directly abuts the 28,000-acre Groton State Forest, a low-use parcel that is a known home to moose, fisher cats, bobcats, and black bear.

Should John and I have been home, the farm's other llamas would have alerted us to the presence of the bear, most likely in time to have saved this ewe. She was valued at approximately $300, and the owner of the ewe was reimbursed for her loss by the State of Vermont. To date, the sheep are never left unguarded and we have since had no reoccurring predator problems.

Actual Experiences with Guardian Llamas

Instance One. A large sheep farm in New York was experiencing 30 to 40 losses per year. This sheep producer tried all types of guard dogs, donkeys, trapping, shooting, even camping out with the flock at night. After the addition of a screened guardian llama, the producer's losses dropped to 5 per year. This particular guardian llama followed the sheep in every night for shelter. At one point, the sheep came in without the llama. The sheep producer searched for the llama and found him guarding a single ewe that was vulnerable and injured with a broken leg. This llama had stayed with the injured animal. Help arrived when his presence was missed by the shepherd.

Instance Two. A newly screened guardian llama was placed with a small sheep producer, replacing a donkey that was chewing on lambs' ears. One day, a 1200-pound bull moose ambled into the pasture. Although it was not a threat to the sheep, the guardian llama "rounded up" all the sheep and ran them into the barn. This amazing llama then stood guard in the gateway between the moose and the sheep until the moose lost interest and wandered away.

Instance Three. A newly placed and screened guardian llama in Michigan was observed actively chasing a member of the canine species from a sheep pasture.

Instance Four. A goat producer experienced a severe storm that took down a good portion of his fencing. Panicking, he started to search for the goats and guardian llama that had found their way out through the damaged fencing. Within an hour, the llama was seen leading the goats down the driveway toward his farm. Neighboring farmers reported that they had seen a llama frantically running around trying to gather up "his" goats.

This guardian llama follows his flock out to pasture every morning.

Gale Birutta

Instance Five. A show horse breeder was having trouble with neighborhood dogs chasing and running his prized breeding stock. With the addition of a llama, the dog attacks ceased and the llama, allowed to free range, patrolled the perimeter of the farm keeping neighborhood dogs off the property completely.

The stories go on and on. It is not uncommon to hear people who have guardian llamas say: "If I hadn't seen it with my own eyes, I would never have believed it!"

Guardian Llamas as a Business

Screening llamas as livestock guardians can be a very lucrative business. With the increasing numbers of llamas, many breeders are seeking homes for extra males.

Locating llamas for a screening program is not especially difficult, but not all llamas are guardians. One first must have the facilities to screen these incoming llamas. They must be kept separate from other llamas, preferably not within sight's distance. Ideally, you would need to house sheep or other livestock on your farm to fairly and intelligently screen these llamas.

Alternatively, you can locate an alternate livestock producer who is willing to work with you through regional or state producer groups. An arrangement can be made whereby he utilizes the unscreened potential guardian llama at no charge, while you in turn utilize his livestock to "set the stage" for an actual guarding situation. Using the checklist on page 279, ask the livestock producer to monitor the llama for you. Keep in mind, he is not a llama breeder nor is he familiar with llamas. You are the expert and only you know how llamas may react in certain situations. Again, the best way to screen llamas as guardians is to keep sheep or other livestock ON YOUR FARM.

Advertising in regional sheep publications and joining local producer groups will give you more than enough exposure. You *must screen* these animals and not just be looking to make quick money. In all instances, compassion is the highest priority, as these producers have most likely experienced severe financial loss by the time they come to you.

Llamas entering a program for screening can come on a consignment basis. After the llama is successfully screened, pay the owner. Llamas can be purchased for a program like this for as little as $300, and can be resold to a producer for predator protection for as much as $2500. Again, be compassionate. Above all, try to help the producer who has suffered severe losses. He may have lost as much as $10,000 in livestock in a matter of several weeks and will be financially tapped out at this point. Offer the producer fair payment terms, and he will return the favor by sending clients to you.

Postscript: To Always Have Llamas

Llamas are tranquil animals and have an amazingly calming effect on humans. Many people who are stressed from their daily jobs find peace in the company of their llamas.

Llamas steal our hearts. We can interact with them without the amount of work that conventional livestock require.

Llamas are a constant source of amusement. At dusk, adults and babies begin their playtime. With a gait much like that of an antelope, they bound on all fours with their feet together, called "sproinging" in the llama world. They will then break gait and run as fast as they can around their pastures. All join in, including the breeding sires and pregnant females.

Llamas love to pose. At shows and fairs, my junior herd sire is always on the alert for television cameras. When we videotape him, he becomes more stately and imperious and actually seems to present his best profile! This same llama, when he was a yearling, rode in the back of John's Suzuki Samurai for four hours to get to the Eastern States Exposition in Massachusetts. On the way back, after four days with 100,000 people, this little guy fell asleep on my shoulder.

Our amazing animals bring joy to our hearts and peace to our minds. Once llamas become part of your life, you will never be without them. They have enriched our lives and will be with us always.

Lithia Llamas

Llamas are the perfect end to a fine day.

GLOSSARY

Alpaca. Domesticated camelid cousin of the llama, native to South America, generally smaller in size.

Anemia. Blood-related medical condition in which the llama has a low red blood cell (RBC) count and low packed cell volume (PCV).

Angular Limb Deformity. A genetic or nutritionally related condition, usually diagnosed in baby llamas, in which the legs are crooked.

Banana ears. The desired long, curved shape of a llama's ears.

Batts. Bundled layers of processed fleece.

Berserk Male Syndrome. Aggressive behavior in a male llama related to (1) prior overhandling by humans while it was a cria, and (2) imprinting of, and identification with, humans instead of other llamas.

Blackleg. This clostridial organism resides in the soil and intestines of an animal, causing muscle bruising, swelling, fever, and even death.

Bloodline. Genetic family history.

Blood-typing. A test to determine a llama's specific blood type for the purpose of revealing its family breeding history.

Body scoring. A hands-on and visual evaluation of a llama's body condition.

Breastbone. Bone that runs vertically and central to the ribs.

Britchings. Mechanisms to hold a saddle frame in place, consisting specifically of the breast collar (preventing backward sliding) and the crupper (preventing forward sliding).

Camelids. The family of animals to which llamas, vicunas, alpacas, and guanacos belong.

Cannon bone. The bone between the hock (knee) and the fetlock (ankle).

CBC. Complete blood count.

Chute. Containment structure in which to hold a llama.

Clostridial organism. Oxygen-hating organism.

Coccidia. A highly contagious parasite that is spread through feces. It can cause diarrhea, anemia, weight loss, and even death.

Colostrum. A dam's first milk for her cria, containing vital antibodies and nutrients.

Conformation. Proper proportionate bone structure.

Conjunctivitis. An infection of the eye, commonly known as "pink-eye."

Corpus luteum. The female reproductive structure that produces progesterone when an egg is fertilized.

Creep feed. A specialized smaller feeder for the cria, used to ensure that the baby llama gets enough food.

Cria. A baby llama.

Crimp. The natural curl grown into the llama's fiber from the follicle.

Crupper. A mechanism fastened underneath the llama's tail and hind end to prevent the saddle frame from sliding forward.

Culling. Removing undesirable males or females from the breeding program.

Curl. Less pronounced than crimp, this is a waviness of the fiber grown in from the follicle.

Dam. A female llama with a cria at her side.

Dystocia. A difficult birth.

Electrolytes. Ions such as sodium, calcium, and phosphorus that float in the bloodstream and help maintain a llama's normal metabolism.

Embryo. An unborn baby llama.

Embryonic absorption. This occurs when the pregnant llama's body absorbs the fetal tissue, thus ending the pregnancy.

Enterotoxemia. A clostridial organism found in soil and a llama's intestinal tract; it can cause diarrhea, weakness, lack of coordination, convulsions, coma, and even death.

Eperythrozoon (EPE). A blood-borne parasite that causes various symptoms of infection, including anemia, fever, weight loss, and weakness.

Fighting teeth. Two upper pairs and one lower pair of sharp teeth in the llama, used by male llamas in competition, and often removed.

Forage. Edible vegetation growing in a pasture.

Foundation females. Those females that a particular breeder started with or first purchased as his beginning stock.

Gelding. A castrated male.

Guanaco. Wild camelid cousin of the domestic llama and alpaca.

Guard hair. Crimpless outercoat of a llama that acts as a moisture barrier.

Guardian llama. A llama specially screened to guard livestock.

Hand breeding. A system of introducing a male and female under halter and lead for the purpose of mating.

Herd sire. The male llama on a farm bred for offspring.

Hock. The "knee" area of a llama.

Host. The animal affected by a disease organism.

Hot manure. Manure that is exceptionally high in nitrogen and tends to burn plants through direct contact.

Hybrid vigor. Genetic soundness and health.

Hyperthermia. Heat stress.

Hypothermia. An abnormally low body temperature, caused by prolonged exposure to the cold.

ILA. International Llama Association.

Implantation failure. A malfunction of the body's physical placement of the fertilized llama egg, resulting in a failed pregnancy.

Imprinting. Specific early handling of a cria by a human.

Inbreeding. Mating with a close relative, such as a parent or sibling.

Intact male. Noncastrated male.

Ivermectin (also known as Ivomec). Medication used to treat many llama infections, especially parasitic infections.

Killed vaccine. A vaccine containing only dead bacterial organisms.

Kush. To lie down.

Leptospirosis. An organism that thrives in wet areas and attacks the llama's kidneys, causing blood in the urine, diarrhea, loss of appetite, and fever.

Lice. An external parasite; infestation by this creature can cause skin problems, anemia, and general restlessness.

Linebreeding. Mating of a male and female llama so their offspring remains closely related to a desired ancestor.

Liver fluke. A parasite that inhabits marshy, wet areas; when ingested by the llama, it invades the liver and causes appetite loss, anemia, and digestive problems.

Loafing shed. An open shelter where llamas can rest, escape inclement weather conditions, or socialize.

Maiden female. A female of breeding age who has not yet been bred.

Malignant edema. A clostridial disease caused by an organism living in the soil; this organism thrives on the contents of a llama's intestines and may cause severe swelling around any wound site if the llama gets cut.

Mash. Soft food mixture, used especially for older llamas that may have lost many teeth.

Mastitis. An infection of the mammary glands.

Matron. A female llama who has previously given birth.

Meconium. A cria's first feces.

Medullated fiber. Hair with a hollow core, such as a llama's hair.

Meningeal worm. Aberrant type of parasite normally found in deer, and often passed along to llamas through deer feces in grazing pasture.

Nasal bots. A rare condition; this parasite enters the llama's nasal cavity and sinuses, causing a runny nose, nasal discharge, rubbing of the nose, and extensive sneezing.

Nematodes. Worms.

Open females. Nonpregnant, breeding-age female llamas.

Outbreeding. Mating within the same general genetic population, but between unrelated animals.

Outcrossing. Mating between animals from completely different gene pools.

Outside breedings. When studs are used for hire for their reproduction services, and other breeders ship their females to the studs' farm for a set fee.

Overconditioning. Feeding animals too much, which causes them to become overweight.

Pacing gait. Movement in which both feet on the same side move forward simultaneously.

Palpation. Physical examination by applying pressure of the hand or fingers to a particular surface of the body to determine the condition of an area or organ.

Panniers. Packs that llamas carry on their backs, similar to large saddlebags.

Parasites. Organisms that feed off of other animals.

Pastern. The part of the llama's foot between the fetlock (ankle) and the toes.

PCV. Packed cell volume; important in determining whether a llama is anemic.

Phenotype. The typical physical characteristics of an animal species.

Picket line. A stake screwed into the ground with a line attached so that a llama may graze.

Prepotency. The ability of the male and female to duplicate themselves as closely as possible in any given offspring.

Prepuce. The sac that holds the penis.

Progesterone. A hormone produced by the corpus luteum when a female llama is pregnant.

Rabies. Contracted from the bite of an infected animal, this disease causes abnormal animal behavior and even death.

RBC. Red blood cell; important in determining whether a llama is anemic.

Resorption. This occurs when an animal's body absorbs some of its own tissue, as when a female llama's body absorbs fetal tissue, thus ending the pregnancy.

Rotational grazing. Breaking grazing areas into separate, smaller areas and rotating more densely stocked animals to utilize forage more efficiently.

Ruminants. Animals, such as llamas, that chew their cud.

Sarcoptic mange. Transmitted by a tiny mite, this external parasitic disease causes hair loss, dandruff, and scabs.

SCS. Soil Conservation Service.

Setting up. Same as squaring.

Sheath. The penile area.

Squaring. Placement of an animal so that it stands squarely, evenly, on all four feet.

Tailset. The positioning of the tail in relation to its attachment to the llama's body.

Tapeworm. A worm that attaches itself to the small intestine of the host llama, causing infection.

TDN. Total digestible nutrients.

Tetanus. Clostridial microorganisms that take refuge in soil and an animal's intestinal tract; they release a toxin that affects the llama's nervous system and causes jaw muscle spasms.

Ticks. These parasites feed on blood, attaching themselves to the skin and causing anemia, loss of appetite, weakness, and infected skin.

Topline. The line that runs from the base of the llama's neck to the top of its tail.

Tubing. Mechanism passed through a cria's mouth and down into its stomach, through which colostrum feedings can be made if the baby does not suck.

Ultrasound. A test in which sound waves are reflected off of internal body structures to create images; used particularly to determine a llama's pregnancy and monitor fetal progress.

Underconditioning. Feeding animals too little, which causes them to become underweight.

Vicuna. Wild camelid cousin of the domestic alpaca and llama.

WBC. White blood cell; important in determining the presence of infection.

Withers. The high part of the llama's back, located between the shoulder blades.

APPENDIX A:
ASSOCIATIONS AND
ORGANIZATIONS

National/International

Alpaca and Llama Show Association
(ALSA)
Box 1189
Lyons, CO 80540
(303) 823-0659

Alpaca Owners and Breeders
Association
1140 Manford Avenue, PO Box 1992
Estes Park, CO 80517-1992
(970) 586-5357;
Fax: (970) 586-6685

Auquenido Association of the Americas
PO Box 1066
Somis, CA 93066
(805) 987-0676

Canadian Llama Association
#2 Notre Dame Crescent
LeDuc, Alberta, T9E 6H8
Canada
(403) 986-9562

Classic 2000 Registry
PO Box 211
Pocatello, ID 83204
(208) 232-6456; (800) 398-0832

Guardian Llama Placement Program
National Headquarters and Research
Center
152 Heath Brook Road
Groton, VT 05046
Phone/Fax: (802) 584-3198

International Llama Association
2755 S. Locust Street, Suite 114
Denver, CO 80222
(303) 756-9004
(Information Hot Line)

International Llama Registry
PO Box 8
Kalispell, MT 59903
(406) 755-3438;
Fax: (406) 755-3439

Llama Association of North America
1800 S. Obenchain Road
Eagle Point, OR 97524
(503) 830-5262

Natural Fiber Producers Association
1890 St. George Road
Danville, CA 94526

Regional

Central Valley Llama Owners
9025 Rodden Road
Oakdale, CA 95361
(209) 847-3474

Gold Country Llama Association
7580 Shelborne Drive
Granite Bay, CA 95746
(916) 791-0793

Golden Plains Llama Association
RR 1, Box 96
Phillipsburg, KS 67661
(913) 543-2598

Greater Appalachian Llama Association
PO Box 6992
Harrisburg, PA 17112-0992
(207) 527-2319

Inland Northwest Llama Association
PO Box 762
Vera Dale, WA 99037
(509) 238-4975

Llama Association of the Mid-Atlantic
States
PO Box 252
Ashland, VA 23005

Llamas of Eastern Oregon
64153 Aspen Road
LaGrande, OR 97850
(503) 963-7595

MARICO (Massachusetts/Rhode
Island/Connecticut)
Llama and Alpaca Association
PO Box 414
East Longmeadow, MA 01028-1404
(413) 525-1538

Mid-Central Llama Association
Route 1, Box 77A
Pawnee, IL 62558
(217) 625-7751

New England Alpaca Owners and
Breeders Association
RR 2, Box 3220
Bowdoinham, ME 04008
(207) 268-2257;
Fax: (207) 268-ALPS

Northern Rockies Chapter ILA
1080 Hodgson Road
Columbia Falls, MT 59912-9027
(406) 752-2994

Ohio River Valley Llama Association
27000 Morris/Salem Road
Circleville, OH 43113
(614) 477-6397

Rocky Mountain Llama and Alpaca
Association
7411 North Road 2 East
Monte Vista, CO 81144
(719) 852-4852

South Central Llama Association
PO Box 163654
Austin, TX 78716
(512) 328-9419

Southern States Llama Association
310 Wilson Cove Road
Swannanoh, NC 28778
(704) 298-5637

Southwest Llama Association
PO Box 2631
Flagstaff, AZ 86003
(520) 526-2883

State Associations

Alabama

Alabama Association of Llama Breeders
and Packers
83 Guy Hood Farm Road
Rainbow City, AL 35906
(205) 442-7183

Alaska

Alaska Chapter ILA
HC 31, Box 5247-A
Wasilla, AK 99687
(907) 376-8472

Arizona

Llama and Alpaca Association of
Arizona
1107 East Lockwood Street
Mesa, AZ 85203
(602) 464-8568

California

California Alpaca Breeders Association
(CALPACA)
9234 Champs De Elysses
Forestville, CA 95436
(707) 887-9462;
Fax: (710) 542-0413

California Chapter ILA
25900 Fairview Avenue
Hayward, CA 94542
(510) 582-3393

Central Coast Llama Association
3397 Badger Road
Santa Ynez, CA 93460
(805) 688-8778

Llama Association of Southern
California
PO Box 876
Norco, CA 91760

Western Llama Association
18208 Cull Canyon Road
Castro Valley, CA 94546

Colorado

Alpaca Breeders of the Rockies
3709 County Road 50
Ft. Collins, CO 80521
(303) 482-0350

Llamas Central Colorado
PO Box 17658
Colorado Springs, CO 80935-7658

Florida

Florida Llama Herd
Parish, FL
(941) 776-1302

Georgia

Georgia Llamas and Alpaca Association
3818 Tibbitts Road
Dallas, GA 30132
(770) 445-2855

Hawaii

All Llamas of Hawaii Association
PO Box 721
Hona Kala, HI 96727

Idaho

Idaho Llama Breeders Network
7626 N. 5th West
Idaho Falls, ID 83401
(208) 524-0330

Western Idaho Llama Association
PO Box 190032
Boise, ID 83719
(208) 888-5262

Indiana

Hoosier Llama Association
380 S. 300 East
Columbia City, IN 46725
(219) 244-7936

Iowa

Iowa Llama Association
10439 Bladensburg Road
Ottumwa, IA 52501
(515) 682-1637

Kentucky

Kentucky Llama and Alpaca Association
1710 Watts Ferry Road
Frankfort, KY 40601
(606) 873-1622

Maine

Maine Llama Association
RR3 Box 2990
Skowhegan, ME 04976
(207) 474-6021

Michigan

Michigan Llama Association
4705 Llama Lane
Pottersville, MI 48876
(517) 645-2719

Minnesota

Llamas of Minnesota
8810 Country Road 136SE
Chatfield, MN 55923
(507) 867-3257

Missouri

Missouri Llama Association
PO Box 467
Chillicothe, MO 64601

Montana

Flathead Valley Llama Club
1150 Whitefish State Road
Kalispell, MT 59901
(406) 257-4883

Northern Rockies Llama Association
5700 Head Drive
Helena, MT 59601
(406) 449-7093

Western Montana Llama Marketing
    Network
PO Box 564
Lolo, MT 59847
(406) 273-2535

Nebraska

Nebraska Llama Association
Route 1, Box 32A
Davey, NE 68336

Nevada

Sierra Nevada Network
1032 Kerry Lane
Gardnerville, NE 89410
(702) 265-3177

New Hampshire

New Hampshire Llama Association
979 Isaac Frye Highway
Wilton, NH 03086
(603) 654-2161

New Mexico

New Mexico Llama Lovers
Star Route 303
Placitas, NM 87403
(505) 867-3442

New York

New York Llama and Alpaca
   Association
Salley's Alley
Denver, NY 12421
(607) 326-4850

North Dakota

Llama and Alpaca Association of
   North Dakota
8911 Highway 32
Forman, ND 58032
(701) 724-3059

Ohio

Great Lakes Alpaca Association
Hinckley, OH 44233-9472
(414) 377-8420

Oklahoma

Oklahoma Llama Association
2124 W. 5th Avenue
Stillwater, OK 74074-2815
(405) 624-8839

Oregon

Central Oregon Llama Association
PO Box 53334
Bend, OR 97708
(800) 241-5262

Columbia Alpaca Breeders Association
21750 SW 65th Avenue
Tualatin, OR 97062
(800) 863-7653

Kalmath Basin Llama Association
PO Box 7894
Kalmath Falls, OR 97602
(503) 882-8143

Umpqua Valley Llama Association
Box 1805
Roseburg, OR 97470
(503) 679-6753

Willamette Valley Llama Association
4526 Roberta Ridge Road South
Salem, OR 97302
(503) 581-5760

Pennsylvania

Pennsylvania Llama and Alpaca
  Association
575 Deer Creek Road
Saxonburg, PA 16056
(412) 898-7229

South Dakota

South Dakota Llama Association
1112 Jackson Boulevard
Rapid City, SD 57702
(605) 787-4149

Tennessee

Tennessee Llama Community
202 Ridge Road
Medina, TN 38355
(901) 783-3036

Utah

Utah Llama Association Chapter ILA
2680 W. 5700 S
Mt. Sterling, UT 84339
(801) 245-3529

Vermont

Vermont Llama and Alpaca Association
RR 1, Box 694B
Stamford, VT 05352-9602
(802) 694-1417

Washington

Alpaca Breeders of Washington
29641 NE Timmer Road
Ridgefield, WA 98642
(360) 887-8563

Llama Owners of Washington State
3430 Pacific Avenue SE,
  Suite A-6, 317
Olympia, WA 98501
(206) 893-5222

Mount Baker Llama Owners
PO Box 28753
Bellingham, WA 98228-0753

Southwest Washington Llama Owners
16609A NE 72nd Avenue
Vancouver, WA 98686
(360) 573-4369

SW Washington Llama Association
20719 NE 68th Street
Vancouver, WA 98682
(206) 254-1157

Wisconsin

Organization of Llama Enthusiasts
W8591 Hunters Road
Hortonville, WI 54944
(414) 779-6351

Wisconsin Organization of Llama Lovers
11308 Sherman Road
Cedarburg, WI 53012
(414) 377-9370

Wyoming

Wyoming Llama Owners Association
6095 Raderville Route, Box 3
Casper, WY 82604
(307) 473-1625

APPENDIX B: SUPPLIERS

Llama and Alpaca Supplies

Alternate Livestock Industries, Inc.
PO Box 115
Lebanon, OR 97355
(541) 451-4347
Wholesale only

Llamas Del Sol
5700 West Lovejoy Road
Perry, MI 48872
(517) 625-4965
General gifts

Quality Llama Products
33217 Bellinger Scale Road
Lebanon, OR 97355
(800) 638-4689
General supplies, gifts

Quarry Hill Llamas and Supplies
PO Box 414
East Longmeadow, MA 10128
(800) 245-2627
General supplies, gifts

Rocky Mountain Llamas
7202 North 45th Street
Longmont, CO 80503
(303) 530-5573
General supplies, gifts

Stevens Llama Tique
Route 4, Box 39
Worthington, MN 56187
(800) 469-5262
General supplies, gifts

Useful Llama Items
3540 76th Street
Caledonia, MI 49316
(800) 635-5262
General supplies, gifts

WXICOF
914 Riske Lane
Wentzville, MO 63385
(314) 828-5100
Books, supplies, gifts

Packing and Outfitting Supplies

American Innovation Marketing, Inc.
3292 South Highway 97
Redmond, OR 97756
(800) 824-4288
PYROMID Outdoor Cooking system

Columbia Crest Llamas
415 Confer Road
Kalama, WA 98625-9602
(206) 673-4477
Aluminum pack frames, panniers

Ecollama/Ecopack
Box 8342
Missoula, MT 59807-8342
(406) 542-1625
Canvas pack saddles

Green Mountain Llamas
HCR 33, Box 43
Townshend, VT 05353
(802) 365-7581
Pack saddle frames, pack baskets

The Llama Connection
PO Box 211
Pocatello, ID 83204
(208) 232-6456; (800) 398-0832
Packing equipment

Mt. Sopris Llamas, Unltd.
0270 County Road 111
Carbondale, CO 81623
(800) 767-7479
Pack saddles, panniers, related equipment

Rolling Rock Llamas
1190 Marshall Road
Boulder, CO 80303-7326
(303) 494-8219
Sew-your-own kits

Sanford Enterprises
Box 343
Bluff, UT 84512
(801) 672-2466
Pack saddles, panniers

Snake River Llamas
7626 N. 5th W. # 406
Idaho Falls, ID 83401-5643
(208) 524-0330

Aluminum crossbuck saddle, panniers
Timberline Llamas Inc.
30361 Rainbow Hill Road
Golden, CO 80401-9710
(303) 526-0092
Pack saddles, panniers

Fencing Suppliers

Kiwi Fence
1145 East Roy Furman Highway
Waynesburg, PA 15370-8070
(412) 627-8158
Specializes in New Zealand Electric Fencing

Mills

Cross Creek Valley Wool Mill
4 Ferguson Street
Avella, PA 15312
(412) 587-3222

Frankenmuth Woolen Mill
570 South Main Street
Frankenmuth, MI 48734
(517) 652-8121

Livestock Insurance

Wilkins Livestock Insurers Inc.
Box 24, RR 1
Geneva, NE 68361
(800) 826-9441

Mineral and Supplement Suppliers

Cache La Poudre Minerals
168 Emerald Mountain Ct.
Livermore, CO 80536
1-800-758-0825

Fastrack
11175 Golf Link Road
Turlock, CA 95380
(209) 632-6891

Kimball Farm Llamas
789 East Broadway
Haverhill, MA 01830
(508) 373-0816

Steelhead Specialty Minerals
North 1212 Washington, Suite 12
Spokane, WA 99201
(800) 367-1534

Stone House Farm
New England Camelid Services
Steve R. Purdy, DVM
RR 2, Box 78, Trebo Road
Chester, VT 05143
(802) 875-4503

Herd Management Software

Black Ink Software
PO Box 1951
Beaverton, OR 97075-1951
(503) 641-8596

Forest Enterprises
Llamatrak
603 South Main Street
Arab, AL 35016
(800) 798-4422

Lamaherd
SWCS
65711 Twin Bridges
Bend, OR 97701
(503) 389-1913

Matrix Digital Media
1046 NW 9th
Corvalis, OR 97330
(541) 754-7042

Rite On
Llama Tracker
PO Box 154
Sisters, OR 97759

Sicon Consulting
PO Box 36019
Regina, Saskatchewan S4S 656
Canada
(306) 584-3211

Fertilizer Processing Suppliers, Packaging, and Equipment

American Packaging Corp.
Consumer and Industrial Bag Division
Grant and Ashton
Philadelphia, PA 19114
(215) 698-4800
Packaging bags

Bag Packaging Corp.
PO Box 68, 1215 Spruce Street
Roselle, NJ 07203-0068
(201) 241-6060
Packaging bags

Immunology Products

Triple J. Farms
23404 Northeast 8th Street
Redmond, WA 98053
(206) 868-6263

Blood-Testing Labs

Cornell University Diagnostic
    Laboratory
New York State College of Veterinary
    Medicine
PO Box 5786
Ithaca, NY 14852
(607) 253-3900

M& M Labs
13615 Wabash Road
Milan, MI 48160
(313) 439-2698

Napa Valley Medical Laboratories
3230 Beard Roa
Napa, CA 94558
(707) 257-2750

Rocky Mountain Laboratories
456 Link Lane
Fort Collins, CO 80524
(303) 221-3116

APPENDIX C:
OTHER RESOURCES

Llama and Alpaca Magazines

The Backcountry Llama
2857 Rose Valley Loop
Kelso, WA 98626

Northeast Kingdom Showcase
152 Heath Brook Road
Groton, VT 05046
(802) 584-3198

Llama Banner
PO Box 1968
Manhattan, KS 66505
(913) 537-0320

Llama Life II
5232 Blenheim Road
Charlottesville, VA 22902
(800) 688-4983

The LAMA Link
Drawer 1995
Kalispell, MT 59903-1995
(406) 752-2569

Llamas: The International Camelid Journal
46 Main Street
Jackson, CA 95642
(209) 223-0469

Spinning and Fiber Publications

The Fiberfest Magazine
PO Box 112, Dept. 18
Hastings, MI 49058
(616) 765-3047

Spin-Off
Interweave Press
201 E. 4th Street
Loveland, CO 80537
(303) 669-7672

Books, Videos, Photography

The Alpaca Book
Clay Press, Inc.
PO Box 100
Herald, CA 95638
(800) 401-LAMA

*Caring for Llamas: A Health
and Management Guide*
Rocky Mountain Llama and
Alpaca Association
PO Box 541
Akron, CO 80720
(303) 345-6632

Homestead Press and Productions
152 Heath Brook Road
Groton, VT 05046
(802) 584-3198
Video production, advertising, newsletters

Juniper Ridge Press
PO Box 1278
Olympia, WA 98507
1-800-869-7342
Books, videos (including "Why Llamas")

Llamas Are the Ultimate
Snake River Llamas
7626 North 5th West
Idaho Falls, ID 83402

Llama's Store
PO Box 100
Herald, CA 95638
(800) 401-5262
A complete list is available upon request

McGinnis Video Productions
c/o Betty Barkman
34190 Ledge Road
Tollhouse, CA 93667
Videos

Susan Jones Ley
8685 Hawick Court N.
Dublin, OH 43017
(614) 889-0629
Photography

Wolfgang Bayer Productions
PO Box 915
Jackson Hole, WY 83001
(307) 733-6590
*Videos (including "The Land of the
Llamas")*

Training Resources

Paul and Betty Barkman
Barkman Animal Enterprises
34190 Ledge Road
Tollhouse, CA 93667
(209) 855-6227

Bobra Goldsmith
Rocky Mountain Llamas
7202 North 45th Street
Longmont, CO 80503
(303) 530-5575

John Mallon Training Clinics
19526 Rancho Ballena Road
Ramona, CA 92065
(800) 594-7254

Marty McGee
Zephyr Farm
4251 Pulver Road
Dundee, NY 14837
(800) 883-2670

ACKNOWLEDGMENTS

The following resources were extremely helpful in reviewing material throughout the production of this book.

New England Camelid Services
Stephen R. Purdy, DVM
RR 2, Box 78, Trebo Road
Chester, VT 05143

River Valley Veterinary Hospital
Walter Cottrell
US Route 5
Newbury, VT 05051

Fred and Earlah Swift
Mariah Alpacas
Groton, VT 05046

Smallgrove Interiorscapes
Route 3, Box 5
Montpelier, VT 05602

Liz Marino
Ivory Pond Farm
PO Box 187
South Egremont, MA 01258-0187

Dr. Steve Herbert
Department of Plant and Soil Sciences
Bowditch Hall
University of Massachusetts
Amherst, MA 01003

Dr. William Franklin
124 Science 2
Department of Ecology
Iowa State University
Ames, IA 50011

Lynn Lenker/Tim Barrus
Lithia Llamas
7 Stone Road
Goshen, MA 01032-960

Nancy Chlarson
Quality Llama Product
33217 Bellinger Scale Road
Lebanon, OR 97355

Wes Holmquist
The Llama Connection
PO Box 211
Pocatello, ID 83204

Lee Delaney, D.V.M
Tunic Road
Shaftsbury, VT 05262

INDEX

Page reference numbers in *italics* indicate illustrations; **bold** indicates charts.

About the Author

In 1987, Gale and John Birutta quit their jobs in New Jersey and purchased a 55-acre farm in Vermont's Northeast Kingdom to raise llamas. They named their llama breeding business *Made in Vermont Llamas*. *Northeast Kingdom Llama Expeditions* is the couple's commercial packing enterprise.

Gale's commitment to her breeding program has paid off: Each new generation of her llamas places Grand and Reserve Champion year after year since 1990. Her farm produces some of the finest and most sought-after breeding stock in the Northeast.

Gale's entrepreneurial spirit created many firsts in the llama industry. Gale was the first nationally to develop a llama manure–based organic fertilizer endorsed by a state department of agriculture. Her fertilizer proudly carries the coveted Vermont Seal of Quality. In 1990, Gale was responsible for the initial contact with the Vermont Department of Agriculture to help develop the Vermont Llama Census, laying the groundwork for the later formation of the Vermont Llama and Alpaca Association.

The highly regarded Guardian Llama Placement Program was initiated and developed by Gale in January 1995. Dedicated to researching and placing only the finest screened guardian llamas, the program has placed its "graduates" with diversified livestock producers with predator problems. Gale works with state extension systems and other llama breeders throughout the country, assisting them to establish programs in their region.

Gale's devotion to the llama industry is evident in her tireless commitment to education and the promotion of llamas.

OTHER STOREY TITLES YOU WILL ENJOY

The Family Cow, by Dirk van Loon. This book contains practical, fully illustrated chapters with accurate information on buying, behavior, nutrition, breeds, handling, feeding, milking, health care, calving, and growing feeding crops. 272 pages. Paperback. ISBN 0-88266-066-7.

The Guilt-Free Dog Owner's Guide: Caring for a Dog When You're Short on Time and Space, by Diana Delmar. Easy-to-read chapters remove the anxieties associated with selecting the right dog, housebreaking, exercise, manners, behavior problems, home hazards, travel, and dog health. 180 pages. Paperback. ISBN 0-88266-575-8.

Keeping Livestock Healthy, by N. Bruce Haynes, D.V.M. Learn to prevent disease in horses, cattle, pigs, goats, and sheep through good nutrition, proper housing, and appropriate care. 352 pages. Paperback. ISBN 1-58017-435-3.

Horse Care for Kids, by Cherry Hill. Beginning with how to match the right animal with the right rider and progressing through feeding, grooming, stabling, health care, safety, and more, this book provides everything a young equestrian wants and needs to know about horses. 128 pages. Paperback. ISBN 1-58017-476-0.

Horse Handling & Grooming: A Step-by-Step Photographic Guide, by Cherry Hill. This book covers every aspect of handling and grooming, from equipment and haltering/tying techniques, to bathing and special care for your horse's mane and tail. 160 pages. Paperback. ISBN 0-88266-956-7.

Horse Health Care: A Step-by-Step Photographic Guide, by Cherry Hill. This book covers every aspect of physical care, as well as some basic first aid techniques, feeding techniques, and the proper protective clothing for your horse. 160 pages. Paperback. ISBN 0-88266-955-9.

Horse Sense: A Complete Guide to Horse Selection & Care, by John J. Mettler, Jr., D.V.M. The basics on selecting, housing, fencing, and feeding a horse, including information on immunizations, dental care, and breeding. 160 pages. Paperback. ISBN 0-88266-545-6.

Safe Horse, Safe Rider: A Young Rider's Guide to Responsible Horsekeeping, by Jessie Haas. Beginning with understanding the horse and ending with competitions, this book includes encouraging ideas for a good working relationship in every chapter. Includes chapters on horse body language, safe pastures and stables, catching, leading and tying, grooming safety, and riding out. 160 pages. Paperback. ISBN 0-88266-700-9.

Small-Scale Pig Raising, by Dirk van Loon. Information on penning and handling, health and nutrition, commercial feeds, breeding, physiology, and butchering. 272 pages. ISBN 0-88266-136-1.

Storey's Guide to Raising Chickens, by Gail Damerow. An informative, friendly book that enables both beginning and experienced chicken owners to successfully raise chickens for eggs, for meat, or as a hobby. Covers selecting a breed, taking care of chicks, producing eggs for eating, raising broilers, feeding, troubleshooting, and much more. 352 pages. Paperback. ISBN 1-58017-325-X.

Storey's Guide to Raising Ducks, by Dave Holderread. For both beginners and experts, this book provides detailed, easy-to-follow instructions for housing, feeding, breeding, and controlling disease in your home duck flock. 288 pages. Paperback. ISBN 1-58017-258-X.

Storey's Guide to Raising Dairy Goats, by Jerry Belanger. For both beginners and experts, this book provides detailed, easy-to-follow instructions for housing, feeding, breeding, and controlling disease in your milk goat herd. 288 pages. Paperback. ISBN 1-58017-259-8.

Storey's Guide to Raising Poultry, by Leonard Mercia. For both beginners and experts, this book provides detailed, easy-to-follow instructions for housing, feeding, breeding, and controlling disease in your poultry. 352 pages. Paperback. ISBN 1-58017-263-6.

Storey's Guide to Raising Rabbits, by Bob Bennett. For both beginners and experts, this book provides detailed, easy-to-follow instructions for housing, feeding, breeding, and controlling disease in your rabbits. 256 pages. Paperback. ISBN 1-58017-260-1.

Storey's Guide to Raising Sheep, by Paula Simmons and Carol Ekarius. For both beginners and experts, this book provides detailed, easy-to-follow instructions for housing, feeding, breeding, and controlling disease in your sheep herd. 400 pages. Paperback. ISBN 1-58017-262-8.

Storey's Guide to Raising Turkeys, by Leonard Mercia. For both beginners and experts, this book provides detailed, easy-to-follow instructions for housing, feeding, breeding, and controlling disease in your turkeys. 208 pages. Paperback. ISBN 1-58017-261-X.

Your Calf: A Kid's Guide to Raising and Showing Beef and Dairy Calves, by Heather Smith Thomas. A friendly and encouraging children's reference book featuring information on calf selection, housing, feeding, health, behavior, and showing. 192 pages. Paperback. ISBN 0-88266-947-8.

Your Chickens: A Kid's Guide to Raising and Showing, by Gail Damerow. A friendly and encouraging children's reference book featuring information on chicken selection, housing, feeding, health, behavior, and showing. 160 pages. Paperback. ISBN 0-88266-823-4.

Your Goats: A Kid's Guide to Raising and Showing, by Gail Damerow. A friendly and encouraging children's reference book featuring information on goat selection, housing, feeding, health, behavior, and showing. 176 pages. Paperback. ISBN 0-88266-825-0.

Your Guinea Pig: A Kid's Guide to Raising and Showing, by Wanda L. Curran. A friendly and encouraging children's reference book featuring information on guinea pig selection, housing, feeding, health, behavior, and showing. 160 pages. Paperback. ISBN 0-88266-889-7.

Your Horse: A Step-by Step Guide to Horse Ownership, by Judy Chapple. Highly readable for all ages and packed with practical information on buying, housing, feeding, training, riding, and handling medical problems. 144 pages. Paperback. ISBN 0-88266-353-4.

Your Puppy, Your Dog: A Kid's Guide to Raising a Happy, Healthy Dog, by Pat Storer. A friendly and encouraging children's reference book featuring information on puppy and dog selection, housing, feeding, health, and behavior. 176 pages. Paperback. ISBN 0-88266-959-1.

Your Rabbit: A Kid's Guide to Raising and Showing, by Nancy Searle. A friendly and encouraging children's reference book featuring information on rabbit selection, housing, feeding, health, behavior, and showing. 160 pages. Paperback. ISBN 0-88266-767-X.

Your Sheep: A Kid's Guide to Raising and Showing, by Paula Simmons and Darrell L. Salsbury, D.V.M. A friendly and encouraging children's reference book featuring information on sheep selection, housing, feeding, health, behavior, and showing. 128 pages. Paperback. ISBN 0-88266-769-6.

*These and other Storey books are available wherever books are sold
and directly from Storey Publishing,
210 MASS MoCA Way, North Adams, MA 01247,
or by calling 1-800-441-5700.
Or visit our Web site at www.storey.com*